Learning from Nature How to Design New
Implantable Biomaterials:
From Biomineralization Fundamentals to
Biomimetic Materials and Processing Routes

NATO Science Series

A Series presenting the results of scientific meetings supported under the NATO Science Programme.

The Series is published by IOS Press, Amsterdam, and Kluwer Academic Publishers in conjunction with the NATO Scientific Affairs Division

Sub-Series

I.   **Life and Behavioural Sciences**   IOS Press
II. **Mathematics, Physics and Chemistry**   Kluwer Academic Publishers
III. **Computer and Systems Science**   IOS Press
IV. **Earth and Environmental Sciences**   Kluwer Academic Publishers
V.   **Science and Technology Policy**   IOS Press

The NATO Science Series continues the series of books published formerly as the NATO ASI Series.

The NATO Science Programme offers support for collaboration in civil science between scientists of countries of the Euro-Atlantic Partnership Council. The types of scientific meeting generally supported are "Advanced Study Institutes" and "Advanced Research Workshops", although other types of meeting are supported from time to time. The NATO Science Series collects together the results of these meetings. The meetings are co-organized bij scientists from NATO countries and scientists from NATO's Partner countries – countries of the CIS and Central and Eastern Europe.

Advanced Study Institutes are high-level tutorial courses offering in-depth study of latest advances in a field.
Advanced Research Workshops are expert meetings aimed at critical assessment of a field, and identification of directions for future action.

As a consequence of the restructuring of the NATO Science Programme in 1999, the NATO Science Series has been re-organised and there are currently Five Sub-series as noted above. Please consult the following web sites for information on previous volumes published in the Series, as well as details of earlier Sub-series.

http://www.nato.int/science
http://www.wkap.nl
http://www.iospress.nl
http://www.wtv-books.de/nato-pco.htm

Series II: Mathematics, Physics and Chemistry – Vol. 171

Learning from Nature How to Design New Implantable Biomaterials:
From Biomineralization Fundamentals to Biomimetic Materials and Processing Routes

edited by

R. L. Reis
Department of Polymer Engineering,
University of Minho,
Braga, Portugal

and

S. Weiner
Department of Structural Biology,
Weizmann Institute of Science,
Rehovot, Israel

Kluwer Academic Publishers

Dordrecht / Boston / London

Published in cooperation with NATO Scientific Affairs Division

Proceedings of the NATO Advanced Study Institute on
Learning from Nature How to Design New Implantable Biomaterials: From
Biomineralization Fundamentals to Biomimetic Materials and Processing Routes
Alvor, Algarve, Portugal
13–24 October 2003

A C.I.P. Catalogue record for this book is available from the Library of Congress.

ISBN 1-4020-2647-1 (PB)
ISBN 1-4020-2644-7 (HB)
ISBN 1-4020-2648-X (e-book)

Published by Kluwer Academic Publishers,
P.O. Box 17, 3300 AA Dordrecht, The Netherlands.

Sold and distributed in North, Central and South America
by Kluwer Academic Publishers,
101 Philip Drive, Norwell, MA 02061, U.S.A.

In all other countries, sold and distributed
by Kluwer Academic Publishers,
P.O. Box 322, 3300 AH Dordrecht, The Netherlands.

Printed on acid-free paper

TABLE OF CONTENTS

CONTRIBUTORS

J. Aizemberg, Bell Laboratories/Lucent Technologies, 600 Mountain Ave., Murray Hill, NJ 07974, USA.

C. M. Alves, 3B's Research Group – Biomaterials, Biodegradables and Biomimetics, Department of Polymer Engineering, University of Minho, Campus de Gualtar, 4710-057 Braga, PORTUGAL.

H. S. Azevedo, 3B's Research Group – Biomaterials, Biodegradables and Biomimetics, Department of Polymer Engineering, University of Minho, Campus de Gualtar, 4710-057 Braga, PORTUGAL.

F. Barrère, iBME, Twente University, PO Box 217, 7500 AE Enschede, THE NETHERLANDS.

B. Ben-Nissan, Department of Chemistry, Materials and Forensic Science, University of Technology, Sidney, PO BOX 123, Broadway 2007, NSW, AUSTRALIA.

R. C. Bielby, Imperial College London, Tissue Engineering and Regenerative Medicine Centre, Faculty of Medicine, Chelsea & Westminter Campus, 369 Fulham Road, London SW10 9NH, UK.

P. Calvert, Department of Textile Sciences, University of Massachusetts at Dartmouth, N. Dartmouth, MA 02747, USA.

P. Ducheyne, Department of Bioengineering, Center for Bioactive Materials and Tissue Engineering, University of Pennsylvania, 3320 Smith Walk, Philadelphia 19104, Pennsylvania, USA.

P. Fratzl, Max-Planck-Institute of Colloids and Interfaces, Department of Biomaterials, Science Park Golm, 14424 Potsdam, GERMANY.

M. E. Gomes, 3B's Research Group – Biomaterials, Biodegradables and Biomimetics, Department of Polymer Engineering, University of Minho, Campus de Gualtar, 4710-057 Braga, PORTUGAL.

K. De Groot, Biomaterials Research Group, Leiden University, Prof. Bonkhorstlaan 10-D, 3723 MB Bilthoven, iBME, Twente University, PO Box 217, 7500 AE Enschede and IsoTis, B.V., Prof. Bronkhorstlaan 10-D, 3723 MB Bilthoven, THE NETHERLANDS.

P. Habibovic, iBME, Twente University, PO Box 217, 7500 AE Enschede, THE NETHERLANDS.

G. Hendler, Natural History Museum of Los Angeles County, Los Angeles, CA 90007, USA.

G. Jabbour, University of Arizona, Tucson AZ 85721, USA.

I. B. Leonor, 3B's Research Group – Biomaterials, Biodegradables and Biomimetics, Department of Polymer Engineering, University of Minho, Campus de Gualtar, 4710-057 Braga, PORTUGAL.

A. L. Oliveira, 3B's Research Group – Biomaterials, Biodegradables and Biomimetics, Department of Polymer Engineering, University of Minho, Campus de Gualtar, 4710-057 Braga, PORTUGAL.

I. Pashkuleva, 3B's Research Group – Biomaterials, Biodegradables and Biomimetics, Department of Polymer Engineering, University of Minho, Campus de Azurém, 4800-058 Guimarães, PORTUGAL.

J. M. Polak, Imperial College London, Tissue Engineering and Regenerative Medicine Centre, Faculty of Medicine, Chelsea & Westminter Campus, 369 Fulham Road, London SW10 9NH, UK.

S. Radin, Department of Bioengineering, Center for Bioactive Materials and Tissue Engineering, University of Pennsylvania, 3320 Smith Walk, Philadelphia 19104, Pennsylvania, USA.

B. D. Ratner, Center of Bioengineering and Department of Chemical Engineering, University of Washington 98195, USA.

R. L. Reis, 3B's Research Group – Biomaterials, Biodegradables and Biomimetics, Department of Polymer Engineering, University of Minho, Campus de Gualtar, 4710-057 Braga, PORTUGAL.

A. J. Salgado, 3B's Research Group – Biomaterials, Biodegradables and Biomimetics, Department of Polymer Engineering, University of Minho, Campus de Gualtar, 4710-057 Braga, PORTUGAL.

S. Weiner, Department of Structural Biology, Weizmann Institute of Science, Rehovot, 76100, ISRAEL.

Y. Yoshioka, University of Arizona, Tucson AZ 85721, USA.

H. Yuan, IsoTis SA, Prof. Bronkhorstlaan 10, 3723 MB. Bilthoven, THE NETHERLANDS.

P. Zaslansky, Department of Structural Biology, Weizmann Institute of Science, Rehovot, 76100, ISRAEL.

PREFACE

The development of materials for any replacement or regeneration application should be based on the thorough understanding of the structure to be substituted. This is true in many fields, but particularly exigent in substitution and regeneration medicine. The demands upon the material properties largely depend on the site of application and the function it has to restore. Ideally, a replacement material should mimic the living tissue from a mechanical, chemical, biological and functional point of view. Of course this is much easier to write down than to implement in clinical practice.

Mineralized tissues such as bones, tooth and shells have attracted, in the last few years, considerable interest as natural anisotropic composite structures with adequate mechanical properties. In fact, Nature is and will continue to be the best materials scientist ever. Who better than nature can design complex structures and control the intricate phenomena (processing routes) that lead to the final shape and structure (from the macro to the nano level) of living creatures? Who can combine biological and physico-chemical mechanisms in such a way that can build ideal structure-properties relationships? Who, else than Nature, can really design smart structural components that respond *in-situ* to exterior stimulus, being able of adapting constantly their microstructure and correspondent properties? In the described philosophy line, mineralized tissues and biomineralization processes are ideal examples to learn-from for the materials scientist of the future.

Typically, the main characteristics of the route by which the mineralized hard tissues are formed is that the organic matrix is laid down first and the inorganic reinforcing phase grows within this organic matrix/template. Bone, tooth, lobster and crabs exoskeletons, oyster shells, coral, ivory, pearls, sea urchin spines, cuttlefish bone, are just a few of the wide variety of biomineralized materials engineered by living creatures. Many of these biological structural materials consist on inorganic minerals combined with organic polymers. The study of these structures has generated a growing awareness that the adaptation of biological processes may lead to significant advances in the controlled fabrication of novel and better-engineered smart-materials. To date, neither the elegance of the biomineral assembly mechanisms nor the rather complex composite microarchitectures could be duplicated by non-biological methods. This is true in spite of the fact that substantial progress has been made in understanding how biomineralization occurs. However, most of this knowledge is yet to be used on relevant industrial applications, namely on the design of appropriate biomimetic routes that will lead to the development of a new generation of implantable biomaterials.

Biomimetics is a new very important field of science that studies how Nature designs, processes and assembles/disassembles molecular building blocks to fabricate high performance mineral-polymer composites (e.g., mollusc shells, bone, tooth) and/or soft materials (e.g., skin, cartilage, tendons) and then applies these designs and processes to engineer new molecules and materials with unique properties. Studies can focus on: the development of methods to reveal the mechanisms through which organic assemblies such as proteins/peptides can determine the biomineral structure of tooth and bone; the determination of tooth and bone hierarchical structure; deciphering of biological basis of biomineralization using paradigms from marine invertebrates; understanding fundamentals of dental enamel proteins self-assembly; the design of a new generation of tooth and bone structure materials based on lessons from the marine mussel and spider silk proteins; and the development of "mineral delivery vehicles" based on phospholipids self-assembly for the repair or re-mineralization of enamel, dentin as well as bone defects. For instances, bone formation is a particularly complex process, on which hydroxylapatite precipitation seems to be associated, initially, with matrix vesicles and subsequently with collagen fibres. Very small crystals are formed, parallel to one another, under the influence of the collagen fibrils structure, as a result of the composition of the inorganic salts present in the body fluids. Bone is synthesized as a complex composite and the organization and interfacial chemistry of the components are optimised for functional use by means of cell-mediated processes.

Furthermore, in the last few years it is becoming well established that the essential requirement for an artificial material to exhibit a bone-bonding behaviour is the formation, on its surface, of a calcium phosphate (Ca-P) similar to bone apatite. In fact, the presence of an apatite like layer on the surface of an orthopaedic biomaterial is considered as a positive biological response from the host tissues. So, it is expected that any material coated with such a bone-like apatite layer might show a bioactive behaviour when implanted. Several works have been trying to learn from the natural processes and to develop synthetic *in-vitro* methodologies, whereby a mineral phase is deposited in a particular polymer matrix or on the surface of a polymeric substrate. In all these works, a very important issue is to generate the chemical conditions at the interface that induce precipitation of mineral phase. Many works have already introduced the term biomimetic (related to the formation of a Ca-P layer on the surface of an implant material) into the biomaterials community.

We have decided to organize a NATO Advanced Study Institute on "*Learning from Nature How to Design New Implantable Biomaterials: From Biomineralization Fundamentals to Biomimetic Materials and Processing Routes*", when we realized that there was a clear need for a course that would address all the above referred to topics. In fact it was our deep believe that there *was* a necessity for a course that addressed, in an integrated way, topics that go from understanding biomineralization processes of different mineralized tissues (that means: not only bone, tooth, etc.) to the use of that science to engineer new biomimetic processes and materials. In fact only an understanding of the relevant fundamentals and a simultaneous application oriented view will lead to the design of new biomimetic materials and processing routes (including production of biomimetic coatings). However the biomineralization and biomaterials research communities have not been working side by side in the past few years. In fact, to our knowledge, no course has addressed before this topic in such an integrated and 'looking forward' perspective. The only meeting on which both communities regularly meet is the Gordon research Conference on Biomineralization. However there is almost no training content on that has most of the talks are aimed at presenting new breakthroughs on biomineralization science to best scientists (most of them well established) in the field. So we thought that an ASI could be a complementary tool as it seemed to us to be the best forum to educate and brainstorming on this area of such strategic importance. This was in our opinion achieved with success. The ASI also helped to integrate the NATO Partner countries in this moving edge technology by means of inducing joint activities and student exchanges with NATO countries Institutions.

The main aims of this Advanced Study Institute were to review, in a tutorial and comprehensive manner, the actual scientific knowledge and recent R&D achievements on biomineralization phenomena, especially those related to organic polymeric matrixes. To understand the fundamentals involved. To comprehend the present state of the art on the use of bioactive ceramics and glasses and other mineralized materials on bone, cartilage and tooth (and other human tissues) regeneration and replacement. To discuss what can be learned from Nature in order to develop new biomaterials. To use the understanding of biomineralization processes on the development of new biomimetic materials and processing routes. To review the present status of research and industrial activities on the important field of biomimetics, and to discuss and brainstorm on its potential evolution. To discuss all the most important areas related to these multidisciplinary fields, trying to help on creating new collaboration and new multidisciplinary hybrid researchers that can also be involved on technology transfer. The main aim is to give the students tools to play a key leading role, guiding research and industrial spin-offs, on the evolution of this particular area in the coming decades. In fact, biomimetics is believed to be, together with tissue engineering one of the most challenging and promising areas on biological driven materials research in the coming decades.

The present book summarizes most of the information delivered during the different lectures and mimics the scientific quality that was presented in the meeting. It is organized in four different sub-topics:
-   Structure and mechanical Functions in Biological Materials
-   Bioceramics, Bioactive Materials and Surface Analysis
-   Biomimetics and Biomimetic Coatings
-   Tissue Engineering of Mineralized Tissues

It is composed by 13 chapters by 10 different groups. The book is mainly what will remain from the NATO-ASI course that was held from the 13th to the 24th of October 2003 in Alvor, Algarve, Portugal. Its structure reflected the integrated and multidisciplinary approach needed in this particular field. The course addressed a wide range of topics, joining together the world-leaders on most of the relevant fields. The Faculty not only gave tutorial lectures, but was always very interactive with the participants trying to open their minds for the future of the field. The lecturers also tried to maximize discussion between themselves and with the participants. The course was organized in several topics and was complemented by short presentations and posters delivered by the participants. The best works presented by the participants have been invited to submit a full manuscript to be considered for publication in a special issue of *Materials Science & Engineering: Part C Biomimetic and Supramolecular Systems* of which I will be the Guest Editor.

Finally, I must say that, as most of you know, nobody can organize a course without the help of hard working people and support from several institutions. I would of course first of all like to thank the NATO Scientific Division for their support that made possible for me to organize another NATO-ASI course and the publication of the present book. I would like also to acknowledge the contributions of my co-director Steve Weiner that really needs no introduction in this field. He was a great support whenever I needed it. The members of the scientific committee and several of lecturers made a lot of useful suggestions. All the invited speakers that accepted our invitation and made the course and the book possible are gratefully acknowledged. The contributions of several of the speakers to this book made possible to produce a state-of-the-art volume to be used by researchers all around the world in the coming years.

But the course, and its program, was also made by the ASI students and their contributions. As said before the best contributions from the ASI students will be published in a special issue of *Materials Science & Engineering: Part C Biomimetic and Supramolecular Systems.* All the supporting institutions, namely the Foundation for Science and Technology of Portugal (FCT), are grateful acknowledged. University of Minho and the Department of Polymer Engineering that have supported me, and my students, in so many ways also deserve a word of appreciation. But I am especially grateful to all of my Post-doc fellows, PhD Students and staff colleagues, not forgetting my personal executive assistant, that work daily on the *3B's Research Group – Biomaterials, Biodegradables and Biomimetics* (www.dep.uminho.pt/3bs) that I have the pleasure of directing. We are now more than 40 people of which only 5 are staff members, all the others are young, bright and ambitious students. The outcome of the ASI was mainly the result of their hard work, devotion, commitment, and of their own ambitions and aspirations. They have put a great number of hours on the enterprise or organizing the course and realized that this was an important organization for all of us. Several of them also supported me on preparing this book. I cannot refer all the names herein, but if you find one of the members of the *3B's Research Group – Biomaterials, Biodegradables and Biomimetics* (that I have the pleasure of directing) in one of the meetings you attend, please just speak with her/him and you will see how fortunate I am for being able to advise such a wonderful group of young and bright researchers!

Please enjoy the science and the lessons contained in this book. We really hope the book will be a useful research and education tool and that it can give the readers the same degree of satisfaction we could experience when preparing it for publication.

Rui L. Reis
(Director of the NATO-ASI Course)

13 - 24 October 2003
Alvor, Algarve, Portugal

1. Structure and Mechanical Functions in Biological Materials

STRUCTURE-MECHANICAL FUNCTION RELATIONS IN BONES AND TEETH

S. WEINER and P. ZASLANSKY
Department of Structural Biology
Weizmann Institute of Science, Rehovot, Israel 76100

1. Introduction

An understanding of the relationships between structure and function in biological materials is fascinating, and may well lead to basic insights into the manner in which they function. Furthermore, it can provide new ideas for developing novel materials. In the case of tissues relevant to humans, such as bones and teeth, this knowledge will certainly improve our ability to provide treatment for pathology or even design better replacement parts, in particular by identifying the key structural properties that need to be built in to the synthetic material.

The problems are however manifold. The structures of many biological tissues, and in particular bones and teeth, are hierarchical and hence very complicated. They are also inhomogeneous, exhibiting a graded or stepped variation in organization even within a specific hierarchical level. It is thus misleading to relate to the structure, and any discussion of structure in relation to function needs to define the hierarchical level being investigated and the variations that occur within this level. Function too is a problematic concept. Whereas individual bones may well perform specific functions, the most common bone type, lamellar bone, appears to be multifunctional – the concrete of the vertebrate skeleton [1]. Thus relating the structure of lamellar bone to function is in essence an effort to understand how it performs its multifunctional role. Teeth on the other hand, do perform specific functional tasks, such as applying compressive forces during mastication. Although the precise manner in which this is done is also complex, the notion of relating tooth structure to function in teeth conceptually at least appears to be appropriate. Furthermore as different teeth have varying structures and morphologies, it is the hope that these may one day be well understood at the materials level in terms of the functions they perform.

Here we will provide brief overviews of the structures of bones and teeth, and then focus on specific topics that are of particular interest. The structure of the paper is thus:

Bone: Brief overview of bone structures.
  The lamellar structure – towards isotropy.
  Functional benefits of remodeling.

R.L. Reis and S. Weiner (eds.),
Learning from Nature How to Design New Implantable Biomaterials, 3-13.
© 2004 *Kluwer Academic Publishers. Printed in the Netherlands.*

Teeth: Brief overview of vertebrate teeth structures.
         Dentin –isotropy achieved?
         The working part of the tooth.

2. Bone

2.1. THE HIERARCHICAL STRUCTURE

Figure 1 summarizes the hierarchical structure of the bone family of materials at least with regard to the major structural types discussed here. This subject is reviewed by Weiner and Wagner [2]. Briefly, the major components of bone comprise thin plate-shaped crystals of carbonated hydroxylapatite, some 50 x 25 x 2 4nm in average dimensions, type I collagen fibrils and water. In most bone, the weight proportions of the 3 major components are 60-70% crystals, 20-30% collagen and other organic components and the remainder is water. With time, the proportion of the crystals increases at the expense of the water. In most bone the proportions of collagen and other matrix components remain constant, although it is possible that with increasing age 1 this decreases as well. Within a given bone there may be a gradient of increasing mineral content from the last-formed surface inwards. Thus the proportions of the major components are variable, and these variations are well known to directly affect the mechanical properties of the bone [3]. As these variations may be localized within the bone, a good understanding of the manner in which they influence the biomechanics remains to be achieved.

The major components are all organized into the basic building block of bone, namely the mineralized collagen fibril. Some of the crystals are located within the fibrils where they are arranged in parallel layers that traverse the fibril cross-section [4], and other crystals particularly in older bone are also located between fibrils [5].

The mineralized fibrils are generally arranged in bundles, and the manner in which these bundles are organized determines the various bone family types that have been recognized. Figure 1 schematically depicts the organization of just 4 such types, parallel fibered bone, woven bone, lamellar bone and bulk dentin. For a more complete description of bone family types, see [6].

Several bone types may be juxtaposed in the same bone. For example bone that is newly formed at the surface is often organized into extended arrays of parallel lamellae. In reptiles and fish this bone type, known as circumferential lamellar bone, usually remains as the major structural type, whereas in most mammals, and certainly in almost all large mammals, circumferential lamellar bone is a transient form that is replaced during the process of remodeling. In many bovid long bones, parallel-fibered bone and lamellar bone types are juxtaposed initially during the formation of the bone. This combination is known as fibrolamellar bone (also called plexiform bone). It too however is usually replaced after some time during the remodeling process by lamellar bone. The structural type that replaces the primarily formed bone is osteonal (also known as Haversian) bone (Fig. 1). This is also comprised of lamellar bone, but unlike circumferential lamellar bone, the lamellae are arranged into elongated cylinders wrapped around a central blood vessel (lacuna). Osteonal bone is the major bone type of humans, and is thus of much biomedical significance. One interesting and much debated issue, is what are the mechanical advantages of osteonal bone over the primary formed bone types – a topic that will be discussed in more detail below. For a listing of the distributions of bone types among different taxa see [7] and associated papers.

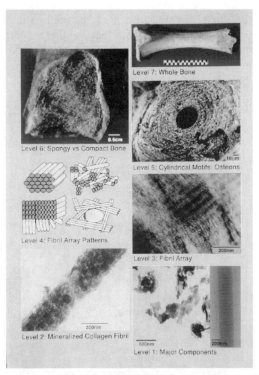

Figure 1. Hierarchical structure of materials in the bone family, following Weiner and Wagner [2]. The bone family of materials includes in addition to the various bone types, mineralized tendon and the different forms of dentin. (Figure is reproduced from Weiner and Wagner [2].

There are many aspects of the complex hierarchical structure of bone that bear a direct relation to its varied mechanical functions. Currey for example has highlighted the importance of the variations in the mineral content [8]. Here I will focus on just two aspects of the structure-mechanical function relations that we and many others have investigated, namely the lamellar structure and the possible benefits of remodeling.

2.2. THE LAMELLAR STRUCTURE

Surveys of the distribution of lamellar bone in different taxonomic groups show that it is present in reptiles, amphibians and is particularly widespread in mammals [7]. It is also present in small and large animals, both terrestrial and aquatic. Thus load-bearing is not a common denominator that accounts for its distribution. In fact lamellar bone is present in load-bearing long bones, as well as shock resistant skull bones. All these parameters are consistent with lamellar bone being able to fulfill many functions [1].

2.2.1. *Structure of the Lamellae.*

Even though this is a subject that is still somewhat controversial [1], the structural model presented here is basically consistent with the original orthogonal plywood model proposed by Gebhart [9], as well as the twisted plywood model proposed [10] in terms of collagen fibril orientations. Figure 2 is a schematic representation of the structure as proposed by [11]. For more details see [12]. The rotated plywood model can be described by two structural rotations. The one rotation is of 5 bundles of mineralized fibrils that are successively oriented in a plane parallel to the lamellar boundary by increments of approximately 30 degrees. Because there appear to be only 5 such bundles and not 6, the structure is asymmetric. The second rotation is of the mineralized fibrils around their fibril axes. This rotation changes the orientations of the layers of crystals within the fibril bundle. This progresses from one side of a lamella to the other.

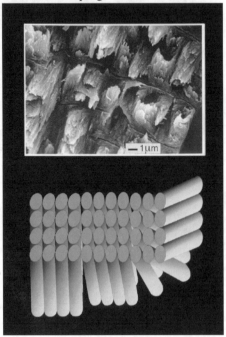

Figure 2. Scanning electron micrograph of the fracture surface of a baboon tibia, showing the different sub-lamellae that constitute an individual lamella (top). Schematic illustration of the 5-layer structure of a lamella, where each cylinder represents a mineralized collagen fibril (not drawn to scale) (bottom). Figure is adapted from Weiner and Wagner [2].

What could be the mechanical benefits of such a complex and rather elegant structure? Ziv *et al.* [13] and Liu *et al.* [14] have performed microhardness measurements of lamellar bone in different orientations and have compared them with those of parallel fibered bone also in different directions. The extent of anisotropy of the lamellar bone is significantly less than that of the parallel fibered bone. As parallel fibered bone is essentially an extended array of the basic building block, the mineralized collagen fibril, this implies that lamellar bone is less anisotropic than the building block from which it is constructed, and hence better able to withstand mechanical challenges in different directions. Thus from this point of view, lamellar bone can function as a multipurpose material. Note however that this refers to the hardness properties. In terms of elastic properties, lamellar bone is also less anisotropic than parallel fibered bone [15].

2.2.2. Benefits of Remodeling.

The remodeling process wherebye ostoclasts remove existing bone and osteoblasts replace it with osteonal bone, results in the bones of most large mature mammals being composed almost entirely of osteonal bone (Fig. 1). Even in small mammals, such as rats, where there is limited remodeling, it preferentially occurs around areas of muscle attachment at the expense of the primary circumferential lamellar bone. So it has not escaped the notice of many investigators that there must be some benefits of osteonal bone as compared to the bone it replaces, which can be woven bone, fibrolamellar bone or as noted circumferential lamellar bone. It is therefore somewhat surprising that primary fibrolamellar bone has a higher tensile strength than remodeled osteonal bone [15,16]. This comparison is however between two different bone types. More pertinent, especially in the case of understanding the benefits of remodeling in human bone, is whether the lamellar bone arranged into osteons has any beneficial mechanical properties compared to extended arrays of circumferential lamellae that are initially deposited? This has however proved particularly difficult to determine, because the primary lamellar bone is usually not available in volumes large enough to be subjected to rigorous mechanical testing in different orientations, so that it can be compared to osteonal bone.

A rare opportunity arose to directly compare secondary osteonal and primary circumferential lamellar bone, when Liu *et al.* [17] noted that in the tibiae of certain growing young adult baboons that were given a rather large dose of the drug alendronate, remodeling had ceased completely, and the entire cross-section of the midshaft of the diaphysis was composed of circumferential lamellae bones. In the control animals the tibiae had remodeled normally to form osteonal bone. Alendronate significantly reduces osteoclast activity [18]. Liu *et al.* [17] thus had an ideal opportunity to compare both the elastic and fracture properties of circumferential lamellar bone and osteonal bone from an age matched group of animals. They made all the measurements in 3 point bending and in 5 different directions, 0, 30, 45, 60 and 90 degrees, with 0 being the direction of the long dimension of the specimen and the long axis of the bone. Surprisingly, after adjusting for the slight differences in mineral content and porosity, it was found that in terms of elastic and fracture properties, the primary and secondary formed bones were essentially identical (Fig.3). Thus there are no benefits of remodeling for these aspects of the materials properties. There were however significant differences in the shapes of the stress-strain curves themselves particularly in the zone where irreversible damage accumulates. There was much more damage related to fracture in the osteonal bone than in the circumferential lamellar bone, and inevitably after massive failure, the two pieces of broken bone were completely separated in circumferential lamellar bone, but never in osteonal bone. Thus clear-cut benefits of remodeling are that the remodeled bone can absorb a lot more fatigue damage, and when fracture finally occurs the two pieces stay together. This must increase the chances of wound healing processes being able to repair the fracture, and thus represent a very significant survival benefit. Perhaps this is one reason why osteonal bone is so prevalent particularly in large active mammals. Furthermore, Currey [19] has pointed out that healed fractures in limb bones of birds and mammals are surprisingly frequent (see Table 10.1 page 311), thus supporting the notion that effective fracture healing is a real benefit for survival.

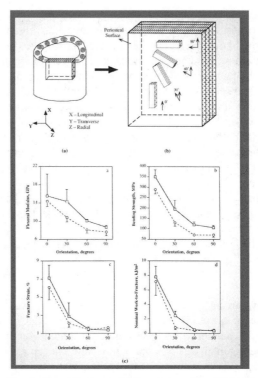

Figure 3. Comparison of the elastic and fracture properties of circumferential lamellar bone (solid lines) and osteonal bone (dashed lines) obtained from the midshafts of baboon tibia. The top row shows the locations of the samples and their alignments relative to the long axis of the bone. Figure is adapted from *Liu et al.* [17].

Another more subtle benefit of osteonal bone was also noticed. The fact that the lamellae are asymmetric in structure and are therefore prone to buckling under compression, is essentially eliminated in osteons as the structure is wrapped into a cylinder. Thus two opposing walls of the cylinder balance each other structurally and the result is an overall symmetrical structure. It is most likely that there are other structural-mechanical benefits of remodeling. For an insightful discussion of the benefits of remodeling see Currey [19], where he has listed the possible benefits as removing dead bone, improving the blood supply, mineral homeostasis, changing the grain and taking out microcracks. It has been shown that the remodeled bone preferentially replaces bone with a high proportion of microcracks [20]. Currey's list of possible benefits is also probably correct. That there is often no single benefit for a given phenomenon, but many, is the way of biology.

3. Teeth

Vertebrate teeth all have a hard wear resistant outer working surface and a bulk which is composed of a more pliant material [21]. There are two types of outer layers, enamel and enameloid. Enamel is composed mainly of long thin spaghetti-shaped crystals of carbonated hydroxyl apatite that are associated with a matrix of proteins (excluding the protein collagen). The crystals are arranged in bundles and the bundles can be oriented in 3 different directions. Enameloid is present only in the teeth of fish. It contains crystals that are similar in shape to enamel, but in many species they are also fluoridated. They however are embedded in a matrix that contains collagen. The thickness of the enamel layer varies, but is usually several hundred microns (reviewed in [22]).

The main body of the tooth is composed of the more pliant material, dentin. Dentin is in fact one of the members of the bone family of materials, as its basic building block is the mineralized collagen fibril and the crystals are also thin and plate-shaped [2] . One significant difference between the organization of the mineralized collagen fibrils in dentin as compared to bone is that in dentin adjacent fibrils in a bundle are rotated with respect to each other and thus the crystal layers are not aligned [23]. In bone they tend to be aligned. The organization of the crystal bundles in dentin is such that they are all basically aligned in the same plane, but within the plane there is little or no preferred orientation. This plane is lies perpendicular to the orientation of one of the hallmarks of dentin: namly the dentin tubules [24]. These tubules sized 1-2 microns in diameter, extend from the pulp cavity almost to the boundary between the enamel and the dentin (Fig. 4) to the so-called dentino-enamel junction (DEJ).

There are a variety of different types of dentin. The two structural types most pertinent to this discussion are peritubular and intertubular dentin, shown in figure 4. In the crown of the tooth, but not in the root, the tubules are surrounded by a layer of relatively dense dentin called peritubular dentin. It is also composed of plate-shaped crystals aligned in layers [26], but there is very little collagen [25]. It has been proposed that peritubular dentin crystals form in a collagenous matrix, but that the latter is subsequently degraded [26]. Intertubular dentin is located between the tubules surrounded by a peritubular envelope (Fig. 4). In root dentin only intertubular dentin is present.

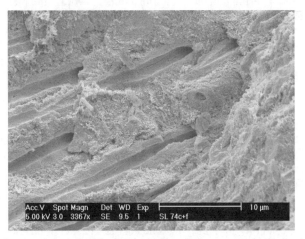

Figure 4. Scanning electron micrograph of the fracture surface of human crown dentin showing the tubules surrounded by dense peritubular dentin. The more fibrous textured material is intertubular dentin.

Even though dentin and bone belong to the same family of materials [2], there are some important differences besides the structural organization. Dentin crystals do not appear to increase in size with age, and the proportions of mineral to matrix remain the same [27]. Furthermore, these proportions vary in a systematic and graded way across the tooth section. This has been well demonstrated by indentation studies, which are sensitive to mineral content [28] and more recently in a diverse and systematic mapping study of crown dentin [29]. It has also been shown that the gradients are different on the buccal and lingual sides of the tooth [30]. It has even been shown that individual peritubular collars have graded hardness properties in cross-section [31], as does enamel [32]. These features are clearly functional, but what exactly they contribute to the mechanical properties of a whole tooth, are not well understood.

3.1. INTERTUBULAR DENTIN – DEGREE OF ANISOTROPY

The degree of anisotropy is a key materials property that may reflect function. A key question is to what extent is dentin anisotropic? An inspection of the structure, with most of the mineralized fibrils arranged in one plane, suggests that it should be as anisotropic as bone.

Wang et al. [23] addressed this issue using microindentation on the roots of human teeth. The roots were used because they comprise mainly intertubular dentin, as opposed to the crowns that are more complicated. By carefully cutting specimens of dentin parallel to the incremental plane, and also along the other two orthogonal planes, Wang et al. [23] obtained the surprising result that in essence intertubular dentin is isotropic at least with respect to the manner in which it responds to indentation. This observation is consistent with other studies, including Kinney et al. [31] who used nanoindentation to study both intertubular and peritubular dentin. Wang et al. [23] suggested that the key structural features that account for this apparent isotropy are the more or less random arrangement of the fibril bundles within the plane perpendicular to the tubules, and most important the fact that adjacent fibrils within a bundle are not aligned azimuthally. Clearly the isotropic nature of the intertubular dentin affords major benefits for tooth function in that the chances of having planes of weakness are minimilaized, and applied stresses can be redistributed in different directions without being unduly influenced by the dentin structure itself.

3.2. WHOLE TOOTH DESIGN FEATURES

The overall strategy of having an outer hard working surface overlying a more pliant material, is certainly a basic design feature of most teeth, among the vertebrates and among many invertebrates. Juxtaposing such different materials and having them work together, is however not trivial. It is known from microhardness profiles across human teeth sections, that the zone below the dentino-enamel junction (DEJ) is a lot softer than the reminder of the crown dentin, and the much harder enamel [23,28,32]. This raised the interesting possibility that this soft zone acts as a gasket or cushion allowing the two different materials to work together. Wang and Weiner [33] used Moiré fringes to map the strain on tooth slices under compression and indeed proved that much of the strain was taken up in this zone. This was subsequently confirmed using Moiré interferometry by Wood et al. [34] at much higher resolution. Thus a 200 micron soft zone in human teeth can indeed be regarded as the working part or one of the working parts of the tooth. Moiré interferometry has also been used to show the strain distribution in human teeth subjected to thermal changes. Interestingly, here too a difference was observed between the buccal and lingual sides [35].

The manner in which the strain is distributed in the soft zone of the tooth slice under compression, is by no means symmetrical. This in turn suggested that we need to understand much more about tooth deformation during mastication in 3 dimensions. With this in mind, Zaslansky *et al.* [36] developed a new methodological approach to this problem combining holographic interferometry with Electronic Speckle Pattern-correlation Interferometry (ESPI). We have designed a means of mapping whole tooth deformation under water with a resolution of tens of nanometers. Performing measurements of teeth, and in fact all biological materials in the hydrated state is essential, as dehydration seriously changes material properties (eg. [37]). Initial results show that whole human teeth do indeed deform asymmetrically. We have also devised a means of measuring the elastic moduli of areas just a few microns squared. Elastic moduli can be measured over even smaller areas using the recovery properties of surface indentations [38]. This approach is however somewhat problematical, as microfracturing can influence the recovery process [32]. ESPI can thus be used to directly map the changes in the elastic modulus across tooth sections. This should provide insights into understanding the advantages of having the mineral content grade in a systematic manner from one side of the tooth to the other.

Much still needs to be understood about whole tooth function. This is of basic importance, and may also have much application in the field of dentistry in terms of designing new materials that are more compatible with the natural dental materials. Furthermore worn tooth reconstructions should also be designed such that the working parts of the tooth are taken into account.

4. Concluding Comment

The more we understand the details of the structure-mechanical function relations of mineralized hard tissues, such as bone and teeth, the more we will be able to utilize this knowledge to design more effective biomaterials and replacement parts for humans. These materials will not necessarily need to be similar in structure and composition to the materials that they are replacing, but should be capable of performing their basic mechanical functions while at the same time be compatible with the biological environment in which they are placed. Achieving this goal still requires a much better understanding of the manner in which these tissues perform their normal functions.

References

1. Weiner, S., Addadi, L. and Wagner, H. (2000) Materials design in biology, *Mat. Sci. Engn. C* 11, 1-8.
2. Weiner, S. and Wagner, H.D. (1998) The material bone: structure- mechanical function relations, *Ann. Rev. Mat. Sci.* 28, 271-298.
3. Currey, J.D. (1984) *The Mechanical Adaptions of Bones*, Princeton University Press, Princeton, New Jersey.

12

4. Weiner, S. and Traub, W. (1986) Organization of hydroxyapatite crystals within collagen fibrils, *FEBS Lett.* **206**, 262-266.
5. Lees, S. and Prostak, K. (1988) The locus of mineral crystallites in bone, *Conn. Tissue Res.* **18**, 41-54.
6. Francillon-Vieillot, H., de Buffrenil, V., Castanet, J., Geraudie, J., Meunier, F.J., Sire, J.Y., Zylberberg, L. and de Ricqles, A. (1990) Microstructure and mineralization of vertebrate skeletal tissues, in J.G. Carter (eds.), *Skeletal Biomineralization. Patterns, Processes and Evolutionary Trends*, van Nostrand, New York, pp. 499-512.
7. Enlow, D. and Brown, S. (1956) A comparative histological study of fossil and recent bone tissues. Part I, *Texas J. Sci.* **8**, 405-443.
8. Currey, J.D. (1984) Effects of differences in mineralization on the mechanical properties of bone, *Philos. Trans. R. Soc. London* **B 304**, 509-518.
9. Gebhardt, W. (1906) Ueber funktionell wichtige Anordnungsweisen der eineren und groberen Bauelemente des Wirbeltierkmochens.II. Spezieller Teil Der Bau der Haversschen Lamellensysteme und seine funktionelle Bedeutung, *Arch. Entwickl. Mech. Org.* **20**, 187-322.
10. Giraud-Guille, M.M. (1988) Twisted plywood architecture of collagen fibrils in human compact bone osteons, *Calcif. Tissue Int.* 167-180.
11. Weiner, S., Arad, T. and Traub, W. (1991) Crystal organization in rat bone lamellae, *FEBS Lett.* **285**, 49-54.
12. Weiner, S., Traub, W. and Wagner, H.D. (1999) Lamellar bone:structure-function relations, *J. Struct. Biol.* **126**, 241-255.
13. Ziv, V., Wagner, H.D. and Weiner, S. (1996) Microstructure-microhardness relations in parallel-fibered and lamellar bone., *Bone* **18**, 417-428.
14. Liu, D., Ziv, V., Wagner, H.D. and Weiner, S. (2000) Mechanical properties - structure relations of parallel fibered and lamellar bone: in-plane anisotropy, *Am. Acad.Orthopaedic Surgeons (in press)*
15. Currey, J. (2002) *Bones*, Princeton University Press, Princeton.
16. Currey, J. (1959) Differences in the tensile strength of bone of different histological types, *J. Anat.* **93**, 87-95.
17. Liu, D., Wagner, H.D. and Weiner, S. (2000) Bending and fracture of compact circumferential and osteonal lamellar bone of the baboon tibia, *J. Mat. Sci. Mat. Med.* **11**, 49-60.
18. Balena, R.T., Toolan, B.C., Shea, M., Markatos, A., A., M., E.R., L., S.C., O., E.E., S., Klien, J.G., Frankenfield, H., Quartuccio, D., Fioravanti, H., Clair, C., Brown, J., Hayes, E., W.C. and G.A., R. (1993) The effectsof a 2-year treatement with the aminobisphosphonate alendronate on bone metabolism, bone histomorphometry, amd bone strength in ovariectomized nonhuman primates., *J. Clin. Invest* **92**, 2577-2586.
19. Currey, J. (2002) *Bones. Structure and Mechanics*, Princeton University Press, Princeton.
20. Mori, S. and Burr, D.B. (1993) Increased intracortical remodeling following fatigue damage, *Bone* **14**, 103-109.
21. Ten Cate, A.R. (1994) *Oral histology: development, structure and function*, Mosby-Year Book, Inc., St. Luis.
22. Lowenstam, H.A. and Weiner, S. (1989) *On Biomineralization*, Oxford University Press, New York.
23. Wang, R.Z. and Weiner, S. (1998) Human root dentin: structural anisotropy and Vickers microhardness isotropy, *Conn. Tissue Res.* **39**, 269-279.
24. Kramer, I.R.H. (1951) The distribution of collagen fibrils in the dentine matrix, *British Dental Journal* **91**, 1-7.
25. Takuma, S. (1960) Electron microscopy of the structure around the dentinal tubule., *J. Dent. Res.* **39**, 873-981.
26. Weiner, S., Veis, A., Beniash, E., Arad, T., Dillon, J., Sabsay, B. and Siddiqui, F. (1999) Peritubular dentin formation: crystal organization and the macromolecular constituents in human teeth, *J. Struct. Biol.* **126**, 27-41.
27. Dalitz, G.P. (1962) Hardness of dentin related to age, *Australian Dental J.* **7**, 463-464.
28. Craig, R. and Peyton, F. (1958) Elastic and mechanical properties of human dentin, *J. Dent. Res.* **37**, 710-718.
29. Tesch, W., Eidelman, N., Roschger, P., Goldenberg, F., Klaushofer, K. and Fratzl, P. (2001) Graded microstructure and mechanical properties of human crown dentin, *Calcif. Tissue Int* **69**, 147-157.
30. Kishen, A., Ramamurty, U. and Asundi, A. (2000) Experimental studies on the nature of property gradients in the human dentine, *J. Biomed. Mater. Res.* **51**, 650-659.
31. Kinney, J., Balooch, M., Marshall, G.W. and Marshall, S.J. (1999) A micromechanics model of the elastic properties of human dentine, *Arch. Oral Biol.* **44**, 813-822.

32. Meredith, N., Sherrif, M., Setchell, D.J. and Swanson, S.A.V. (1996) Measurement of the microhardness and Young's modulus of human enamel and dentine using an indentation technique, *Archs. oral Biol.* **41**, 539-545.

33. Wang, R. and Weiner, S. (1998) Strain-structure relations in human teeth using Moire fringes, *J. Biomech* **31**, 135-141.

34. Wood, J.D., Wang, R.Z., Weiner, S. and Pashley, D.H. (2003) Mapping of tooth deformation caused by moisture change using Moire interferometry, *Dental Materials* **19**, 159-166.

35. Kishen, A. and Asundi, A. (2001) Investigations of thermal property gradients in the human dentine, *J. Biomed. Mater. Res.* **55**, 121-130.

36. Zaslansky, P., Friesem, A. and Weiner, S. (2004) Measuring the elastic modulus of root dentin in compression and under water using Electronic Speckle Pattern-correlation Interferometry (ESPI), *In preparation*

37. Kahler, B., Swain, M.V. and Moule, A. (2003) Fracture-toughening mechanisms responsible for differences in work to fracture of hydrated and dehydrated dentine., *J. Biomech.* **36**, 229-237.

38. Marshall, D.B., Noma, T. and Evans, A.G. (1982) A simple method for determining elastic-modulus to hardness ratios using Knoop indentation measurements., *Comm. Am. Ceram. Soc.* **65**, 175-176.

HIERARCHICAL STRUCTURE AND MECHANICAL ADAPTATION OF BIOLOGICAL MATERIALS

PETER FRATZL
Max-Planck-Institute of Colloids and Interfaces
Department of Biomaterials
Science Park Golm, 14424 Potsdam, Germany
E-mail: fratzl@mpikg-golm.mpg.de

1. Introduction

Biological tissues such as wood, bone or tooth are hierarchically structured to provide maximum strength with a minimum of material. Many of these materials are cellular solids (e.g., cancellous bone or wood) or gradient materials (e.g., dentin). At the lowest level of hierarchy (that is, in the nanometer range), they are usually fibre composites. Due to this hierarchical structure, there is a variety of different possible designs by changing the arrangement of the components at different size levels. In the case of bone, for example, the variability at the nanometer level is in the shape and size of mineral particles, at the micron level in the arrangement of mineralised collagen fibres into lamellar structures and beyond in the inner architecture, the porosity and the shape of the bone. The mechanical properties of bone are well known to depend strongly on all these parameters [1,2]. A selection of textbooks on the relation between form, hierarchical structure and mechanical properties is given in the references [2-7]. Moreover, on may refer to a number a recent review articles addressing the problem in general [8-12], and treating wood [13-15] or bone [1,16-20], in particular.

The hierarchical structure requires special methods of investigation, since the scales at the nano-, micro- and millimetre level all need to be considered, as well as the interplay between the hierarchical levels. Furthermore, it is essential not just to study the structures statically but during deformation. "Seeing" the structures deform (by microscopic or diffraction methods, for instance) is the key to understand deformation mechanisms. Typically, a whole range of techniques are being used and combined in this context, two of them using x-rays (and in particular synchrotron radiation) are described in section 2. Section 3 reviews some results on bone and dentin, while section 4 focuses on wood.

R.L. Reis and S. Weiner (eds.),
Learning from Nature How to Design New Implantable Biomaterials, 15-34.

2. Studying Hierarchical Biological Materials with (Synchrotron) X-Ray Methods

2.1. SCANNING MICRODIFFRACTION AND SMALL-ANGLE SCATTERING

One of the difficulties in studying hierarchical structures is that many orders of magnitude need to be covered in the structural analysis. This generally implies the necessity to combine many techniques. For instance transmission electron microscopy is sensitive to structures in the range of nanometers, scanning electron microscopy reveals micron and sub-micron structures, while light microscopy may cover the larger scales. X-ray or neutron diffraction methods are mostly sensitive to the nanometer scale. A powerful approach to study hierarchical materials at different scales simultaneously is scanning x-ray diffraction [21] or scanning small-angle x-ray scattering [22]. The principle of this technique is shown in Fig. 1.

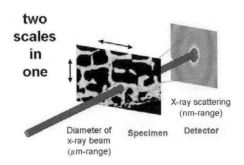

Figure 1. Sketch of a scanning diffraction experiment. Structural information in the nanometer range is obtained from the evaluation of the diffraction (or scattering) pattern. The micrometer range is covered by scanning the specimen across the narrow x-ray beam.

A thin section of the material is scanned across a narrow x-ray beam. The diameter of the x-ray beam defines the lateral resolution of the scanning procedure. It is in the order of 100 micrometer on a laboratory x-ray source [22] and in the order of 1 micrometer (or even below) at synchrotron sources [21,23]. Ideally, the thickness of the specimen should be the same as the beam diameter, d. Then the scattering volume for each individual measurement will be about d^3. Depending on the type of measurement – x-ray diffraction or small-angle scattering - the evaluation of the scattering patterns yields structural data on the nanocomposite, within each d^3-volume separately.

Such local information from x-ray scattering can be advantageously combined with local information from other techniques, e.g. microspectroscopy or nanoindentation on the same specimen. The advantage of such a combination is that information about structure at a certain position of the specimen (determined, e.g., by scanning x-ray diffraction) can be correlated with the mechanical properties (determined, e.g., by nanoindentation) or with the chemical composition (determined, e.g., by light or x-ray spectroscopy) at the very same location inside the material. An example for this approach is discussed in sect. 3.2 for the investigation of the dentin-enamel interface [24].

2.2. IN-SITU DEFORMATION STUDIES WITH SYNCHROTRON RADIATION

In order to uncover the mechanisms of deformation of complex materials, such as those created by Nature, it is advantageous to watch structural changes occurring at the different hierarchical levels during the deformation process. This implies that the actual macroscopic deformation has to be monitored when load is applied, but also requires methods to study deformation at the lower levels simultaneously. One approach is certainly to extract important structural elements, such as individual cells [25] from the wood tissue or individual osteons (that is, a blood vessel surrounded by bone material) from compact bone [26], and study them separately. This requires skilful preparation of the specimens since wood cells or osteons are rather small objects. Moreover, the loading pattern applied to these small elements can hardly be the same as the one applied to them while they are still part of the intact tissue.

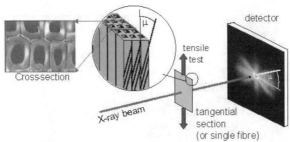

Figure 2. Sketch of an in-situ diffraction experiment (using a wood foil). Structural information on the deformation of the cell walls is obtained by x-ray diffraction (which enables to measure the cellulose microfibril angle, see sect. 4). In this way it is possible to measure the deformation of cell walls which are still integrated in an intact wood tissue [27].

For this reason, a very promising strategy is to use in-situ x-ray diffraction (or small-angle scattering) methods which allow – in some cases – to study the deformation of smaller elements inside an intact tissue which is being deformed. The principle is that the tissue specimen is deformed inside the x-ray beam and the deformation in the elements of the tissue is monitored by the diffraction signal (see Fig. 2). This method will work in those cases where the structural elements (such as fibres, for instance) provide a diffraction signal which depends on the elongation of the fibre. Recent examples where the method has been applied successfully to biological materials are studies of the deformation mechanisms in the wood cell wall [27] (see sect. 4.3) or of the elongation of collagen fibrils within a tendon [28].

3. Hierarchical Structure of Calcified Tissue, Such as Bone and Dentin

3.1. BONE TISSUE

As already mentioned, the mechanical properties of calcified tissues cannot be understood without taking into account all structure levels [17,18]. Some anatomical features known to be important for the strength of the tissue are given on a dimensional scale in Fig. 3.

Typically, a vertebra contains a spongy interior (spongiosa) which prevents it from collapsing when loaded under compression. The struts of this spongy structure, called trabeculae, have a thickness in the order of about 200 microns. In between the trabeculae there is bone marrow. The material of which these trabeculae are made, is lamellar bone. The lamellar structure consists of parallel fibres in each layer, with some rotation of fibre direction in successive layers [18,29,30]. The fibres are collagen fibrils reinforced with calcium-phosphate mineral particles of only a few nanometre thickness [31-34]. The mineral is non-stoichiometric, carbonated hydroxyapatite (dahllite).

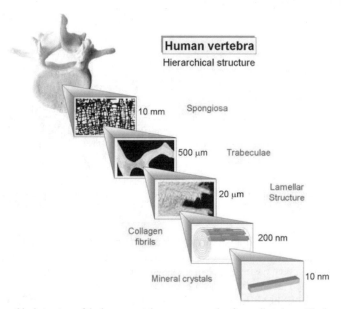

Figure 3. Hierarchical structure of the human vertebra, as an example of cancellous bone. The human vertebra is filled with a highly porous structure of spongy appearance. The trabeculae forming this cellular material are typically 200 micrometer wide and made of a composite of collagen and calcium phosphate mineral. This composite has a lamellar structure, where each lamella consists of layers of parallel mineralised collagen fibrils. Individual collagen fibrils have a diameter of a few hundred nanometers, while the individual reinforcing mineral particles have a thickness of only a few nanometers.

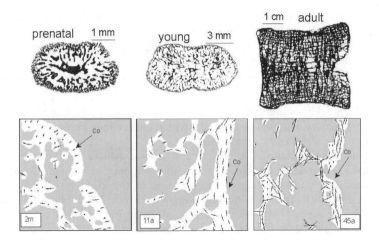

Figure 4. Evolution of the trabecular structure (top row) and the orientation of the collagen-mineral composite (bottom row) in human vertebra, from the prenatal (left) to the adult (right) state [35]. The bottom row shows a small fraction of the vertebra, with cancellous bone to the left of the corticalis (labelled Co). The white areas in the bottom row correspond to bone, while the grey areas indicate the marrow space or the outside of the vertebra. The boxes give the age of the individual (2 months, 11 and 45 years). The direction of the bars indicates the predominant orientation of the elongated (plate-like) mineral nano-particles, measured by scanning-SAXS (see Fig. 1). The length of the bars indicates the degree of alignment. It is evident from the length of the bars that the mineral particles in the collagen tissue are arranged in a much more organized fashion in adult than in young or prenatal bone. Moreover, the orientation inside the cortical shell also changes with age [35].

Even though the list of hierarchical levels is certainly not complete, it is obvious that the structures are spread over at least eight orders of magnitude. Clearly, no single technology can cover such a wide range. While all structures down to about a micrometer in size are accessible to light microscopy, higher resolution can be achieved by other probes, such as scanning- or transmission-electron microscopy, x-ray diffraction (XRD) or small-angle x-ray scattering (SAXS) and, finally, by a variety of spectroscopic techniques, e.g., nuclear magnetic resonance or Fourier-transform infrared spectroscopy. One of the consequences of the hierarchical architecture is that structures at one size-level may vary systematically throughout the tissue on a larger scale.

Some examples are given in the following. Fig. 4 (top row) shows the evolution of the architecture of the spongiosa in the interior of human vertebra [36]. In the embryonal vertebra, the trabeculae are typically arranged in a concentric fashion. In the adult vertebra, on the contrary, the orientation of the trabeculae is mostly vertical and horizontal. This most likely reflects some adaptation to the predominant loading of the vertebra in compression. After birth (see picture labelled "young" in Fig. 4) the architecture changes from the prenatal to adult state. Most interestingly, the changes at the level of the trabeculae are accompanied by changes also in the arrangement of the mineral particles as shown in the bottom row of Fig. 4. The mineral particles, which are known to be parallel to the collagen fibril direction [32,37], typically also follow the struts in trabecular bone. What is even more remarkable is the orientation of the mineral particles in the cortical shell of the vertebra (labelled Co): While the particles are oriented parallel to the outer surface of the vertebra in adults, they are turned 90° to this direction in embryonal vertebra (bottom row, left). The orientation of mineral particles was measured by scanning-SAXS, the technique discussed in section 2. The diameter of the x-ray beam was 0.2 mm, which also corresponded to the thickness of the bone section. The predominant orientation of the mineral particles and their degree of alignment (indicated by bars in the bottom row of Fig. 4) therefore correspond to averages within volumes of the size $0.2 \times 0.2 \times 0.2$ mm^3. Hence, a very short bar (such as in bone of very young individuals) means that – within the volume of investigation – almost all orientations occurred with similar probability.

Given the hierarchical nature of the structure, it is clear that the result of scanning-SAXS measurements depends quite crucially on the scanning resolution chosen for the measurement. Hence, it is extremely important to adapt the spatial resolution to the actual needs for a given structural element. Fig. 5 shows scanning-SAXS data, collected within compact bone, using an x-ray beam diameter of 20µm (that is, with a ten times better scanning resolution than in Fig. 4). The results are superimposed onto an image collected with backscattered electrons [21] and they clearly show the concentric lamellar arrangement around an osteon (that is, a blood vessel surrounded by bone material). With a resolution of 200µm (large dotted circle in Fig. 5), the structure could not have been revealed. Within a circle of this size around the center of the osteon, practically all particle orientations occur, and no specific orientation would remain in the average.

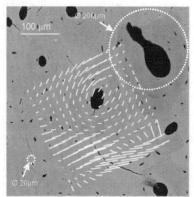

Figure 5. Orientation of mineral particle around an osteon in human compact bone (from [21]). The black ellipse in the centre is the trace of a blood vessel and there are concentric layers of bone lamellae around it, forming the osteon. Several osteons are visible on the backscattered electron image (BEI). The bars are results from scanning-SAXS, obtained at the synchrotron and superimposed on the BEI. They indicate the orientation of mineral platelets with the same convention as in Fig. 4. The specimen thickness and the diameter of the x-ray beam were 20 micrometers in this case. For comparison, a circle with the radius of 200 µm is also shown in the figure (dotted line).

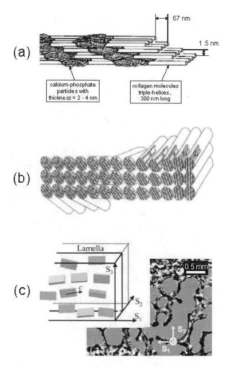

Figure 6. (a) Arrangement of molecules and mineral particles in the collagen fibril, according to [38]. Rotated plywood arrangement of mineralised collagen fibrils in cortical lamellar bone, according to [30]. Orientation of mineral particles in trabeculae of human cancellous bone, according to [39]. MFA (deg)MFA (deg)MFA (deg).

As sketched in Fig. 3, the basic building block of bone materials is the mineralized collagen fibril. Fig. 6a shows a schematic view of the arrangement of mineral particles in fibrils, as measured by transmission electron microscopy [32,38,40]. The particles are very thin (in the order of a few nanometers only) and arranged parallel to the collagen molecules. Obviously, this building block is very anisotropic. Lamellar bone, such as found in osteons, for example (see Fig. 5) is a very common feature in compact bone [29,30,41-45]. Fig. 6b shows a proposal for the arrangement of mineralized fibrils in lamellar bone [29]. Within each lamella, the fibrils run parallel to each other, and the fibril direction is slightly rotated from one such layer to the next, in the form of a rotated plywood structure. A slightly different arrangement was found in the trabeculae of cancellous bone by position-resolved x-ray pole-figure analysis [39]: The mineral particles follow a predominant direction and the structure is, therefore, closer to a fibre texture than to a lamellar arrangement. The predominant direction of the collagen fibrils is lying within the plane of the trabeculum (see Fig. 6c). Further types of arrangement are possible within parallel fibered or plexiform bone (found, e.g., in some types of bovine bone) [41,45,46]. This clearly shows that the elementary building block, the mineralized collagen fibril, is used in many different ways to build higher hierarchical structures in bone. This is most likely related to the different mechanical functions of different bones.

3.2. GRADED STRUCTURE OF DENTIN

The structure of dentin (the bone-like body of teeth) has been studied by scanning-SAXS [24,47] and other methods, such as electron microscopy and nanoindentation [48-50]. A gradient from the enamel towards the root was found both for the structure and for the mechanical properties [24]. Fig. 7 shows the orientation and the thickness of mineral particles in dentin as a function of position. The figure shows that the T-parameter (which is a measure of the thickness of mineral particles) increases systematically from the enamel towards the root. The same section was also investigated by nano-indentation in an atomic force microscope, providing the elastic modulus of the tissue as a function of position. Care was taken to avoid the tubuli (small hollow conducts) in dentin and their immediate surroundings which are known to be slightly overmineralized [24,48]. The indentation modulus (of intertubular dentin) also exhibited a gradient from the enamel towards the root and was plotted in Fig. 7 against the particle thickness (as measured by the T-parameter) determined at the same position on the specimen. The excellent correlation between these two parameters is also shown in Fig. 7. This provides some insight on how the mechanical properties of mineralised tissue was tuned by the type of reinforcing mineral particles. The probable reason for the grading of properties is a better long-term stability against failure of the tooth. Indeed, cracks originating from the enamel seem to stop at the dentin-enamel junction, where the stiffness is lowest [24].

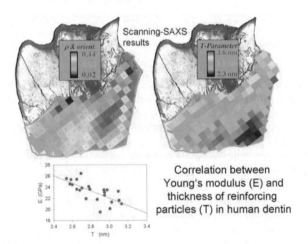

Figure 7. Thickness (right) and orientation (left) of mineral particles in human dentin (from [24]). The T-parameter, which is a measure for the particle thickness, varies from 2.3 to 3.6 nanometers. The degree of alignment ρ (a parameter which is =1, if all the mineral platelets are parallel, and =0, if they are randomly oriented) is larger further away from the enamel (top). The thickness of mineral particles correlates well with the elastic modulus measured (in position-resolved way) on the same specimen by nanoindentation [24].

3.3. MECHANICAL BEHAVIOUR OF THE COLLAGEN MINERAL NANOCOMPOSITE IN BONE OR DENTIN

Virtually all stiff biological materials are composites with components mostly in the size-range of nanometers. In some cases (plants or insect cuticles, for example), a polymeric matrix is reinforced by stiff polymer fibres, such as cellulose or keratin. Even stiffer structures are obtained when a (fibrous) polymeric matrix is reinforced by hard particles, such as carbonated hydroxyapatite in the case of bone or dentin. The general mechanical performance of these composites is quite remarkable. In particular, they combine two properties which are usually quite contradictory, but essential for the function of these materials. Bones, for example, need to be stiff to prevent bending and buckling, but they must also be tough since they should not break catastrophically even when the load exceeds the normal range. How well these two conditions are fulfilled, becomes obvious in the (schematic) Ashby-map in Fig. 8. Proteins (collagen in the case of bone and dentin) are tough but not very stiff. Mineral, on the contrary, is stiff but not very tough. It is obvious from Fig. 8 that bone and dentin combine the good properties of both.

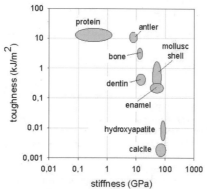

Figure 8. Stiffness and toughness of proteins and mineral (hydroxyapatite and calcite), as well as a few natural protein-mineral composites. The schematic Ashby chart is based on a data compilation by Ashby et al. [9].

It is interesting to notice that the thickness of the mineral particles in bone and dentin is in the order of 2-4 nm only. It is likely that the small size of the components is important for the mechanical performance of these materials. A further important aspect is the very anisometric shape of the mineral particles and their detailed arrangement inside the organic collagen-rich matrix. Indeed, particles are known to be plate-shaped and arranged more or less in parallel with the collagen fibrils [17]. A simple mechanical model based on these structural principles (see Fig. 9) has been studied recently [51,52]. The main feature of this model is that plate-shaped stiff particles are connected by thin layers of soft matrix which is loaded predominantly under shear. A simple expression can be derived for the elastic modulus of the composite which predicts, first, a nearly quadratic dependence on the mineral content (at least for mineral volume fractions smaller than about 60%) and, secondly, a stiffer composite when particles get more anisometric (see Fig. 9). When Young's modulus E is estimated for this model composite, it turns out that the prediction is just intermediate between the Reuss and the Voigt models [2], which represent extreme situations in composites. The predicted dependence of E on the volume fraction of mineral is qualitatively (though not quantitatively) similar to other models for composites as discussed by Akiva et al. [41], for instance.

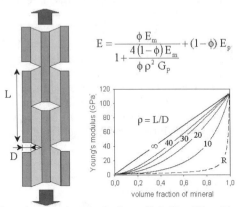

$$E = \frac{\phi\, E_m}{1 + \dfrac{4\,(1-\phi)\,E_m}{\phi\,\rho^2\, G_p}} + (1-\phi)\, E_p$$

Figure 9. Model for the deformation of a mineralized collagen fibril. The stiff particles are platelets (viewed edge-on in the left of the figure) with an anisometric shape defined by the aspect ratio $\rho = L/D$. The matrix between the stiff particles is predominantly shear loaded. An analytical expression for Young's modulus E can be derived by simple considerations [51, 52] and is given in the figure. The graph below the equation shows the dependence of E on the volume fraction of mineral φ, for different values of the aspect ratio ρ as indicated. The broken line (denoted R) corresponds to the standard prediction of the Reuss model for composites [2]. The situation where $\rho \to \infty$ corresponds to the Voigt model for composites [2]. In order to plot E as a function of φ, numerical values of Em =114 Gpa, Ep = 1.5 Gpa, Gp = 1 Gpa have been taken for Young's modulus of the mineral and for Young`s and shear modulus of collagen, respectively.

The model shown in Fig. 9 depends solely on continuum elasticity and as such does not have a characteristic length scale. The stiffness depends on the anisotropy of the particle shape, that is on the ratio $\rho = L/D$, but not directly on the particle thickness: whether D is in the order of nanometers or micrometers does not change the equation for the elastic modulus, at any given value of ρ. Given the construction of the model, most of the tensile load is carried by the hard particles (which provides the stiffness of the composite). This also means that failure of the composite will depend crucially on defects present in the hard and brittle particles. This problem has been analyzed recently by a numerical method which combines finite element analysis and molecular dynamics [52]. It was shown that the susceptibility to flaws in the particles decreases markedly when the particle size in the order of nanometers. The reason is that – in agreement with Griffith's law – the stress concentration at the tip of a small defect depends on the size of this defect. The smaller the defect, the smaller the stress concentration. In fact it could be shown that below some critical size of several nanometers the stress concentration vanishes totally and the the composite becomes defect tolerant [52]. This could explain why nanocomposites such as bone and dentin are quite tough, while they would be brittle with much larger particles. Hence, nano-sized particles seem essential to reduce the inherent brittleness of the composite, but this effect alone cannot fully explain the high toughness of bone. Energy dissipating mechanisms at larger scales are also contributing to prevent cracks from propagating catastrophically. The fibrous and lamellar character of the bone tissue, its anisotropy and inhomogeneity are important factors controlling toughness. A model describing fracture toughness as a function of these various contributions from different hierarchical levels is, however, still missing.

4. Hierarchical Structure of Wood

At the macroscopic level, wood can be considered as a cellular solid, mainly composed of parallel hollow tubes, the wood cells. As an example, the hierarchical structure of spruce wood is shown in Fig. 10. The wood cells are clearly visible in Fig. 10a and they have a thicker cell wall in latewood (LW) than in earlywood (EW), within each annual ring. The cell wall is a fibre composite made cellulose microfibrils embedded into a matrix of hemicelluloses and lignin [53].

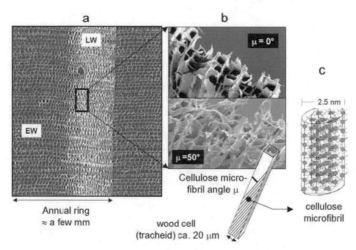

Figure 10. Hierarchical structure of spruce wood. (a) is a cross-section through the stem showing earlywood (EW) and latewood (LW) within an annual ring (from [54]). Latewood is denser than earlywood because the cell walls are thicker. The breadth of the annual rings varies widely depending on climatic conditions during each particular year. (b) shows scanning electron microscopic pictures of fracture surfaces of spruce wood with two different microfibril angles (from [55]). One of the wood cells (tracheids) is drawn schematically showing the definition of the microfibril angle between the spiralling cellulose fibrils and the tracheid axis. (c) is a sketch of the (crystalline part) of a cellulose microfibril in spruce (from [56]).

The cellulose fibrils are wound around the tube-like wood cells with an angle called the microfibril angle (MFA, see Fig. 10). The details of how the fibril direction is distributed in a cell has been investigated by scanning-XRD. Results of these investigations are shown in Fig. 11. An-ray beam of 2 μm diameter was used and diffraction patterns from the cell cross-section were determined in steps of 2 μm over several adjacent cells [57]. X-ray patterns turned out to be anisotropic and even asymmetric due to the non-standard diffraction geometry (Fig. 11, left). This asymmetry could be used to determine the direction of the cellulose fibrils quantitatively. An arrow corresponding to the projection of the unit vector following the fibril direction is shown in Fig. 11 (right) at each point where a diffraction pattern was collected. It is clearly visible that cellulose fibrils in each of the adjacent cells run according to a right-handed helix. The spatial resolution of this experiment was such that only the main cell wall layer (called S2) was imaged. This cell wall layer is enclosed by other, much thinner layers with different cellulose orientation [53].

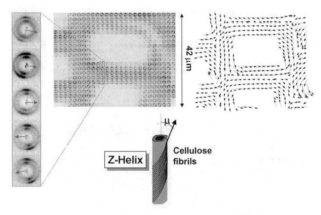

Figure 11. X-ray microdiffraction experiment with a 2 μm thick section of spruce wood embedded in resin (from [57]). Left: typical XRD-patterns from the crystalline part of the cellulose fibrils. Each pattern has been taken with a 2μm wide x-ray beam (at the European Synchrotron Radiation Source). The diffraction patterns are drawn side by side as they were measured. They reproduce several wood cells in cross-section. Note the asymmetry of the patterns in the enlargement (far left) which can be used to determine the local orientation of cellulose fibrils in the cell wall (arrows) [57]. The arrows are plotted in the right image with the convention that they represent the projection of a vector parallel to the fibrils onto the plane of the cross-section. The picture clearly shows that all cells are right-handed helices.

The MFA determines to a large extent the elastic modulus and the fracture strain of wood: When the MFA varies from 0 to 50°, the elastic modulus decreases by about an order of magnitude and the fracture strain increases by a similar factor [55,58]. In some ways, the wood cell behaves like an elastic spring because the stiff cellulose fibrils are wound helically. The steeper the winding angle is, the stiffer wood becomes. This property can be used by the tree to vary considerably the local mechanical properties by growing cells with different microfibril angle. Hence, with the possibilities given by the hierarchical structure, a growing tree can built graded properties into the stem or the branch, according to needs which may change during its lifetime.

4.1. MICROFIBRIL ANGLE DISTRIBUTION IN THE STEM

The first example is the distribution of microfibril angles in the stem. For softwood species (such as spruce or pine) and to some extend also for hardwoods (such as oak), the MFA decreases in older trees from a large value in the pith (about 40°) to very small values closer to the bark [54], see Fig. 12. Since the stem thickens by apposition of annual rings at the exterior, the history of a tree is recorded in the succession of annual rings. Hence, the observation that the microfibril angle decreases from pith to bark indicates that younger trees are optimised for flexibility, while the stem becomes more and more optimised for bending stiffness when the tree gets older (see Fig. 12). A possible explanation for this change in strategy could be a compromise between resistance against buckling (which needs stiffness) and flexibility in bending [54] to resist fracture.

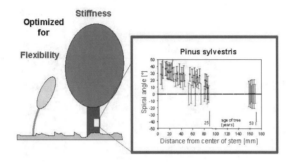

Figure 12. Variation of microfibril angle in a pine stem [54]. Typically, the angle is large in young trees which are, therefore, optimised for flexibility (since large microfibril angles correspond to low stiffness and large extensibility [58]). Older trees are optimised for stiffness.

4.2. STRUCTURAL GRADING IN THE BRANCH

A further example of the grading of material properties by the use of different microfibril angles is the branch. Fig. 13 (left) shows light microscopic pictures of the cell shape on the upper and on the lower side of the branch. Clearly, the lower side (called compression wood) has rounded cells which are quite different from the square-like cell shapes on the upper side (opposite wood). In addition, the microfibril angles are also quite different on the upper and on the lower side, as revealed by scanning-SAXS (Fig. 13, right). The streaks observed in the SAXS patterns are perpendicular to the direction of the cellulose fibrils and reveal, therefore, the microfibril angle [59]. The differences in both micro-and nanostructure between the upper and the lower side are most certainly due to their different mechanical function, the upper side being mostly under tension and the lower side mostly under compression.

A more detailed picture of the distribution of microfibril angles in a branch of spruce wood

28

is shown in Fig. 14. Since the history of growth is stored in the succession of annual rings, it is possible to plot the fibril angle distribution for different ages of the tree. It is remarkable that the young branch (8 years ago) is composed predominantly of flexible wood (with large microfibril angle). With increasing length of the branch (and presumably also an increasing weight from the leaves), a stiff region develops on the upper side of the branch, while the lower side remains with a large MFA (called compression wood) [60]. This means that the asymmetry in the loading pattern is reflected in an asymmetry of the cell microstructure. What is more, the asymmetry develops as a function of age, apparently as a response to modified loading patterns, as the weight and the length of the branch increase. Moreover, it is quite well-known that compression wood my form even in the stem, if there is an asymmetric loading, e.g. due to strong winds from one side.

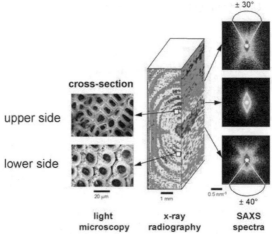

Figure 13. SAXS-patterns from a spruce branch (from [59]). Left: light-microscopic images of cross-sections. Note the round cell shape on the compression side (lower side) of the branch. Centre: X-ray transmission micrograph showing the annual rings. Right: SAXS-patterns showing an MFA of 30° on the upper side and an MFA of 40° on the lower side of the branch.

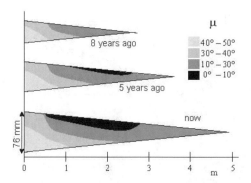

Figure 14. Distribution of microfibril angles measured in a branch of spruce. The age-evolution was deduced from the pattern of annual rings [60].

4.3. DEFORMATION MECHANISM OF THE WOOD CELL WALL

The deformation behaviour of plant cells is quite intricate, particularly at large deformations [61,62]. A typical feature of the stress-strain curve is that a fairly stiff behaviour at low strains is followed by a much "softer" behaviour at large strains (corresponding to a steep increase followed by a smaller slope of the stress-strain curve, Fig. 15c). The mechanisms underlying this deformation behaviour have been studied recently by the diffraction of synchrotron radiation during deformation [27]. Some results of this investigation are shown in Fig. 15. First, the microfibril angle was found to decrease continuously with the applied strain. This relation between micro-fibril angle and strain turned out to be independent of the stress at any given strain. This is shown by stress relaxation experiments (visible as spikes in Fig. 15c), where both strain and microfibril angle μ stay constant, while the stress varies.

30

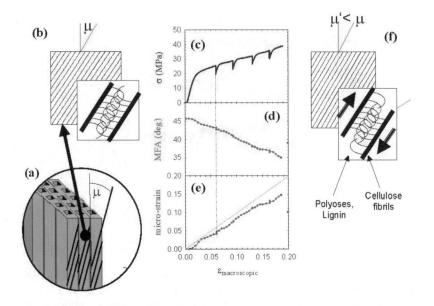

Figure 15. In-situ x-ray diffraction investigation [27] of the deformation of the wood cell wall inside an intact wood section (compression wood of spruce), shown schematically in (a). The dominant cell-wall layer (b) contains cellulose micro-fibrils tilted with the micro-fibril angle μ. Between the micro-fibrils, there is a matrix of hemicelluloses and lignin. (c) shows the stress-strain curve during the deformation experiment. The spikes in the graph correspond to stress relaxation experiments, where the elongation was kept constant. (d) shows the change in micro-fibril angle during the elongation of the specimen. A micro-strain (e) is calculated under the assumption that the cellulose fibrils are rigid and all the deformation is just a tilting of the fibrils and shearing of the matrix in-between (f). The nearly one-to-one correspondence (e) of micro- and macro-strain shows that this is, indeed, the principal mechanism of elongation and that the cellulose fibrils themselves stretch only very little.

In the simplest possible picture, the decrease of the microfibril angle is related to a deformation of each wood cell in a way similar to a spring: The spiral angle of the cellulose microfibrils (that is, the microfibril angle) is reduced from μ to some smaller value μ' and the matrix in-between the fibrils is sheared (Fig. 15b and 15f). In fact, if it is assumed that the elongation of the cellulose microfibrils is negligible, then the elongation of the cell depends solely on the reduction of the microfibril angle as:

$$\text{micro-strain} = \delta(\cos \mu) / \cos \mu = -\tan \mu \; \delta\mu.$$

This expression is plotted in Fig. 15e as a function of the measured macroscopic elongation ($\varepsilon_{macroscopic}$) of the wood tissue. The graph shows that the wood cells actually extend like an elastic spring, and the fact that the cellulose fibrils are not totally inextensible accounts for the slight deviation between the measured data and the straight line in Fig. 15e.

There is, however, one major difference between the behaviour of the wood cell and an elastic spring: indeed – beyond the change in slope in Fig. 15c – the deformation becomes partially irreversible, but without a serious damage to the material [27,62]. The model which can be inferred from the synchrotron diffraction data in Fig. 15 is as follows: When the cell elongates, the microfibril angle is decreasing and the matrix between the cellulose fibrils is sheared. This corresponds to the initial stiff behaviour of the wood cells (initial slope in Fig. 15c). Beyond a certain critical strain, the matrix is sheared to an extent, where bonds are broken and the shearing becomes irreversible. Since some of the bonds are broken, the response is now "softer". After releasing the stress, the unspecific bonds in the matrix reform immediately (a bit like in a velcro connection) and the cell is arrested in the elongated position. In such a model, the matrix is not irreversibly damaged even though the cell is irreversibly elongated [27].

5. Conclusions

Most stiff biological tissues are hierarchically structured, bone and wood being prominent examples. This means that the nanometer structure varies on a micrometer scale. As a consequence, the mechanical properties can be adjusted locally by the organism. Functional gradients and complex structural elements are common in natural tissues. The stiffness of dentin, for example, is graded in such a way that a minimum appears right at the dentin-enamel junction, which is important to prevent catastrophic failure (section 3.2). The flexibility of the material in a branch is also graded to account for the asymmetric loading due to gravitational forces (section 4.2). Mechanical adaptation also leads to age-related changes in the hierarchical structure, both in bone (section 3.1) and wood (sections 4.1 and 4.2). Continued research on natural hierarchical structures is necessary, not only to improve our understanding of biological tissues but also to reveal the strategies and mechanisms used by nature and which may be applied in an engineering context for improving material properties.

The author thanks all collaborators involved during the last few years in the reported studies, including in particular: I. Burgert, J. Färber, H. Gupta, K. Klaushofer, H. Lichtenegger, M. Müller, O. Paris, Ch. Riekel, A. Reiterer, P. Roschger, S. Stanzl-Tschegg, W. Tesch, R. Weinkamer and I. Zizak.

References

1. Currey, J.D. (2003) How well are bones designed to resist fracture?, *J. Bone Miner. Res.* **18**, 591-598.
2. Currey, J.D. (2002) *Bones - Structure and Mechanics*. Princeton University Press, Princeton.
3. Thompson, A.W. (1992) *On Growth and Form - the complete revised edition*. (unaltered republication of Cambridge Univ. Press, 1942), Dover Publications.
4. Mattheck, C., Kubler, H. (1995) *The internal Optimization of Trees*. Springer Verlag, Berlin.
5. Vincent, J.F.V. (1990) *Structural Biomaterials*. revised edition ed. Princeton University Press, Princeton.
6. Wainwright, S.A., Biggs, W.D., Currey, J.D., Gosline, J.M. (1982) *Mechanical Design in Organisms*. Princeton University Press.
7. Niklas, K.J. (1992) *Plant Biomechanics: an engineering approach to plant form and function*. University of

32

Chicago Press, Chicago.

8. Jeronimidis, G. (2000), in M. Elices (eds.), *Structural Biological Materials, Design and Structure-Property Relationships*, Pergamon, Amsterdam, pp. 3-29.

9. Ashby, M.F., Gibson, L.J., Wegst, U., and Olive, R. (1995) The Mechanical-Properties of Natural Materials .1. Material Property Charts, *Proceedings of the Royal Society of London Series a-Mathematical and Physical Sciences* **450**, 123-140.

10. Gibson, L.J., Ashby, M.F., Karam, G.N., Wegst, U., and Shercliff, H.R. (1995) The Mechanical-Properties of Natural Materials .2. Microstructures for Mechanical Efficiency, *Proceedings of the Royal Society of London Series a-Mathematical and Physical Sciences* **450**, 141-162.

11. Jeronimidis, G., and Atkins, A.G. (1995) Mechanics of Biological-Materials and Structures - Natures Lessons for the Engineer, *Proceedings of the Institution of Mechanical Engineers Part C-Journal of Mechanical Engineering Science* **209**, 221-235.

12. Weiner, S., Addadi, L., and Wagner, H.D. (2000) Materials design in biology, *Materials Science & Engineering C-Biomimetic and Supramolecular Systems* **11**, 1-8.

13. Fratzl, P. (2003) Cellulose and collagen: from fibres to tissues, *Curr. Opin. Coll. Interf. Sci.* **8**, 32-39.

14. Mattheck, C., and Bethge, K. (1998) The structural optimization of trees, *Naturwissenschaften* **85**, 1-10.

15. Vincent, J.F.V. (1999) From cellulose to cell, *Journal of Experimental Biology* **202**, 3263-3268.

16. Weiner, S., and Traub, W. (1992) Bone-Structure - from Angstroms to Microns, *Faseb. J.* **6**, 879-885.

17. Weiner, S., and Wagner, H.D. (1998) The material bone: Structure mechanical function relations, *Annual Review of Materials Science* **28**, 271-298.

18. Rho, J.Y., Kuhn-Spearing, L., and Zioupos, P. (1998) Mechanical properties and the hierarchical structure of bone, *Medical Engineering & Physics* **20**, 92-102.

19. Currey, J.D. (1999) The design of mineralised hard tissues for their mechanical functions, *Journal of Experimental Biology* **202**, 3285-3294.

20. Mann, S., and Weiner, S. (1999) Biomineralization: Structural questions at all length scales, *J. Struct. Biol.* **126**, 179-181.

21. Paris, O., Zizak, I., Lichtenegger, H., Roschger, P., Klaushofer, K., and Fratzl, P. (2000) Analysis of the hierarchical structure of biological tissues by scanning X-ray scattering using a micro-beam, *Cell Mol. Biol.* **46**, 993-1004.

22. Fratzl, P., Jakob, H.F., Rinnerthaler, S., Roschger, P., and Klaushofer, K. (1997) Position-resolved small-angle X-ray scattering of complex biological materials, *J. Appl. Cryst.* **30**, 765-769.

23. Riekel, C., Burghammer, M., and Müller, M. (2000) Microbeam small-angle scattering experiments and their combination with microdiffraction, *J. Appl. Cryst.* **33**, 421-423.

24. Tesch, W., Eidelman, N., Roschger, P., Goldenberg, F., Klaushofer, K., and Fratzl, P. (2001) Graded microstructure and mechanical properties of human crown dentin, *Calcif. Tissue Int.* **69**, 147-157.

25. Burgert, I., Keckes, J., Fruhmann, K., Fratzl, P., and Tschegg, S.E. (2002) A comparison of two techniques for wood fibre isolation evaluation by tensile tests on single fibres with different microfibril angle, *Plant. Biol.* **4**, 9-12.

26. Ascenzi, A., Benvenuti, A., Bigi, A., Foresti, E., Koch, M.H.J., Mango, F., Ripamonti, A., and Roveri, N. (1998) X-ray diffraction on cyclically loaded osteons, *Calcif. Tissue Int.* **62**, 266-273.

27. Keckes, J., Burgert, I., Frühmann, K., Müller, M., Kölln, K., Hamilton, M., Burghammer, M., Roth, S.V., Stanzl-Tschegg, S., and Fratzl, P. (2003) Cell-wall recovery after irreversible deformation, *Nature Materials* (in press).

28. Puxkandl, R., Zizak, I., Paris, O., Keckes, J., Tesch, W., Bernstorff, S., Purslow, P., and Fratzl, P. (2002) Viscoelastic properties of collagen: synchrotron radiation investigations and structural model, *Phil. Trans. Roy Soc. Lond. B* **357**, 191-197.

29. Weiner, S., Arad, T., Sabanay, I., and Traub, W. (1997) Rotated plywood structure of primary lamellar bone in the rat: Orientations of the collagen fibril arrays, *Bone* **20**, 509-514.

30. Weiner, S., Traub, W., and Wagner, H.D. (1999) Lamellar bone: Structure-function relations, *J. Struct. Biol.* **126**, 241-255.

31. Landis, W.J. (1996) Mineral characterization in calcifying tissues: Atomic, molecular and macromolecular perspectives, *Conn. Tissue Res.* **35**, 1-8.

32. Rubin, M.A., Jasiuk, I., Taylor, J., Rubin, J., Ganey, T., and P., A.R. (2003) TEM analysis of the nanostructure of normal and osteoporotic human trabecular bone, *Bone* **37**, 270-282.

33. Fratzl, P., Fratzlzelman, N., Klaushofer, K., Vogl, G., and Koller, K. (1991) Nucleation and Growth of Mineral

Crystals in Bone Studied by Small-Angle X-Ray-Scattering, *Calcif. Tissue Int.* **48**, 407-413.

34. Fratzl, P., Groschner, M., Vogl, G., Plenk, H., Eschberger, J., Fratzlzelman, N., Koller, K., and Klaushofer, K. (1992) Mineral Crystals in Calcified Tissues - a Comparative-Study by Saxs, *J. Bone Miner. Res.* **7**, 329-334.

35. Roschger, P., Grabner, B.M., Rinnerthaler, S., Tesch, W., Kneissel, M., Berzlanovich, A., Klaushofer, K., and Fratzl, P. (2001) Structural development of the mineralized tissue in the human L4 vertebral body, *J. Struct. Biol.* **136**, 126-136.

36. Roschger, P., Rinnerthaler, S., Yates, J., Rodan, G.A., Fratzl, P., and Klaushofer, K. (2001) Alendronate increases degree and uniformity of mineralization in cancellous bone and decreases the porosity in cortical bone of osteoporotic women, *Bone* **29**, 185-191.

37. Rinnerthaler, S., Roschger, P., Jakob, H.F., Nader, A., Klaushofer, K., and Fratzl, P. (1999) Scanning small angle X-ray scattering analysis of human bone sections, *Calcif. Tissue Int.* **64**, 422-429.

38. Landis, W.J., Librizzi, J.J., Dunn, M.G., and Silver, F.H. (1995) A Study of the Relationship between Mineral-Content and Mechanical-Properties of Turkey Gastrocnemius Tendon, *J. Bone Miner. Res.* **10**, 859-867.

39. Jaschouz, D., Paris, O., Roschger, P., Hwang, H.S., and Fratzl, P. (2003) Pole figure analysis of mineral nanoparticle orientation in individual trabecula of human vertebral bone, *J. Appl. Cryst.* **36**, 494-498.

40. Landis, W.J., Hodgens, K.J., Arena, J., Song, M.J., and McEwen, B.F. (1996) Structural relations between collagen and mineral in bone as determined by high voltage electron microscopic tomography, *Microscopy Research and Technique* **33**, 192-202.

41. Akiva, U., Wagner, H.D., and Weiner, S. (1998) Modelling the three-dimensional elastic constants of parallel-fibred and lamellar bone, *J. Mater. Sci.* **33**, 1497-1509.

42. Hoffler, C.E., Moore, K.E., Kozloff, K., Zysset, P.K., Brown, M.B., and Goldstein, S.A. (2000) Heterogeneity of bone lamellar-level elastic moduli, *Bone* **26**, 603-609.

43. Liu, D.M., Weiner, S., and Wagner, H.D. (1999) Anisotropic mechanical properties of lamellar bone using miniature cantilever bending specimens, *J. Biomech.* **32**, 647-654.

44. Rho, J.Y., Zioupos, P., Currey, J.D., and Pharr, G.M. (1999) Variations in the individual thick lamellar properties within osteons by nanoindentation, *Bone* **25**, 295-300.

45. Ziv, V., Wagner, H.D., and Weiner, S. (1996) Microstructure-microhardness relations in parallel-fibered and lamellar bone, *Bone* **18**, 417-428.

46. Sasaki, N., Ikawa, T., and Fukuda, A. (1991) Orientation of Mineral in Bovine Bone and the Anisotropic Mechanical-Properties of Plexiform Bone, *J. Biomech.* **24**, 57-61.

47. Kinney, J.H., Pople, J.A., Marshall, G.W., and Marshall, S.J. (2001) Collagen orientation and crystallite size in human dentin: a small angle X-ray scattering study., *Calcified Tissue International* **69**, 31-37.

48. Beniash, E., Traub, W., Veis, A., and Weiner, S. (2000) A transmission electron microscope study using vitrified ice sections of predentin: Structural changes in the dentin collagenous matrix prior to mineralization, *J. Struct. Biol.* **132**, 212-225.

49. Habelitz, S., Balooch, M., Marshall, S.J., Balooch, G., and Marshall, G.W. (2002) In situ atomic force microscopy of partially demineralized human dentin collagen fibrils, *J. Struct. Biol.* **138**, 227-236.

50. Nalla, R.K., Kinney, J.H., and Ritchie, R.O. (2003) Effect of orientation on the in vitro fracture toughness of dentin: the role of toughening mechanisms, *Biomaterials* **24**, 3955-3968.

51. Jager, I., and Fratzl, P. (2000) Mineralized collagen fibrils: A mechanical model with a staggered arrangement of mineral particles, *Biophys. J.* **79**, 1737-1746.

52. Gao, H.J., Ji, B.H., Jager, I.L., Arzt, E., and Fratzl, P. (2003) Materials become insensitive to flaws at nanoscale: Lessons from nature, *Proc. Natl. Acad. Sci. USA* **100**, 5597-5600.

53. Fengel, D., Wegener, G. (1989) *Wood: chemistry, ultrustructure, reactions.* de Gruyter, Berlin.

54. Lichtenegger, H., Reiterer, A., Stanzl-Tschegg, S.E., and Fratzl, P. (1999) Variation of cellulose microfibril angles in softwoods and hardwoods - A possible strategy of mechanical optimization, *J. Struct. Biol.* **128**, 257-269.

55. Reiterer, A., Lichtenegger, H., Fratzl, P., and Stanzl-Tschegg, S.E. (2001) Deformation and energy absorption of wood cell walls with different nanostructure under tensile loading, *J. Mater. Sci.* **36**, 4681-4686.

56. Jakob, H.F., Fengel, D., Tschegg, S.E., and Fratzl, P. (1995) The elementary cellulose fibril in Picea abies: Comparison of transmission electron microscopy, small-angle X-ray scattering, and wide-angle X-ray scattering results, *Macromolecules* **28**, 8782-8787.

57. Lichtenegger, H., Muller, M., Paris, O., Riekel, C., and Fratzl, P. (1999) Imaging of the helical arrangement of cellulose fibrils in wood by synchrotron X-ray microdiffraction, *J. Appl. Cryst.* **32**, 1127-1133.

58. Reiterer, A., Lichtenegger, H., Tschegg, S., and Fratzl, P. (1999) Experimental evidence for a mechanical

function of the cellulose microfibril angle in wood cell walls, *Phil. Mag. A* **79**, 2173-2184.

59. Reiterer, A., Jakob, H.F., Stanzl-Tschegg, S.E., and Fratzl, P. (1998) Spiral angle of elementary cellulose fibrils in cell walls of Picea abies determined by small-angle X-ray scattering, *Wood Sci. Tech.* **32**, 335-345.

60. Farber, J., Lichtenegger, H.C., Reiterer, A., Stanzl-Tschegg, S., and Fratzl, P. (2001) Cellulose microfibril angles in a spruce branch and mechanical implications, *J Mater. Sci.* **36**, 5087-5092.

61. Spatz, H.C., Kohler, L., and Niklas, K.J. (1999) Mechanical behaviour of plant tissues: Composite materials or structures?, *Journal of Experimental Biology* **202**, 3269-3272.

62. Kohler, L., and Spatz, H.C. (2002) Micromechanics of plant tissues beyond the linear-elastic range, *Planta* **215**, 33-40.

2. Bioceramics, Bioactive Materials and Surface Analysis

CALCIUM PHOSPHATE BIOMATERIALS: AN OVERVIEW

HUIPIN YUAN[1], KLAAS DE GROOT[1,2]
[1]Isotis SA, Bilthoven, The Netherlands
[2]Leiden University, Leiden, The Netherlands

Abstract

Calcium phosphates are used by our body to build bones and are being applied to produce biomaterials for bone repair. It is well-known that calcium phosphate biomaterials guide new bone formation, form a tight bond with the newly formed bone, and are therefore, by definition, osteoconductive. Besides their osteoconductive property, it was found that calcium phosphate biomaterials, only with specific physicochemical properties, induce bone formation in non-osseous sites and therefore are osteoinductive. A summary of calcium phosphate biomaterials from osteoconduction to osteoinduction is given in this overview.

Keywords: calcium phosphates, osteoconduction, osteoinduction, bone repair

1. Bone and Bone Repair

Calcium phosphate biomaterials are always discussed in relation with bone repair as calcium phosphate is the main inorganic component of bone.

Although the shape of bone varies in different parts of the body, the physico-chemical structure of bone for these different shapes is basically similar. Biochemically, bone is defined by its special blend of organic matrix (35%) and inorganic elements (65%) [1,2]. The inorganic matter of bone consists mainly of calcium phosphate, significant amounts of citrate and carbonate ions and traces of fluoride, magnesium and sodium. The calcium phosphate in bone is very similar, but not identical, to mineral hydroxyapatite (HA), while the organic matter is largely collagen. At a nano-scale, calcium phosphate deposits on collagen matrix in bone. The mixture of mineral hydroxyapatite and collagen matrix gives bone strength and hardness [1,2].

Despite its strength and hardness, bone is a dynamic living tissue. On the one hand, bone consists of living cells that build bone, namely osteocytes and osteoblasts, while on the other hand, bone has living cells of osteoclasts that resorb old bone during bone remodeling. With both osteoblasts and osteoclasts, bone is constantly being renewed and reconstructed throughout the lifetime of the individual [1-2].

Bone is one of few human tissues that can regenerate themselves. For instance, the fracture of bone activates osteogenesis and ultimately the fracture heals without any

R.L. Reis and S. Weiner (eds.),
Learning from Nature How to Design New Implantable Biomaterials, 37-57.
© 2004 *Kluwer Academic Publishers. Printed in the Netherlands.*

scar formation [1]. Although bone has its own ability to repair, the ability decreases with age, is affected by diseases and other factors, and is limited to small bone defects. Thus far, grafts are necessary to assist bone repair when bone loss is too big, for example in cases of excision of bone tumour, bone sarcoma or other bone diseases, in cases of bone loss in accidents, and in complicated fractured bones that cannot repair themselves.

Several grafts can be the choice in clinics for bone repair [3-11]. Autograft, which is harvested from the patients' own body, has no problem of biocompatibility and immunoresponse. It contains osteogenic cells and osteoinductive growth factors (such as bone morphogenetic proteins, BMPs), allows rapid bone regeneration and reconstruction of bone loss, and is therefore still the "golden standard" and the most-used bone graft. However, autograft is limited in amount, needs a secondary operation, and causes pain and morbidity. Moreover it is not always successful in all cases. Allograft, which is provide by the donors, has osteogenic cells and osteoinductive growth factors as autograft does, but its osteogenic capacity may decrease due to its immunologic rejection, meanwhile the supply of allograft is limited and there is the risk of disease transmission. Demineralised bone matrix (DBM), produced from donor bone and containing osteoinductive growth factors, may help bone repair, but its osteogenic capacity is not always guaranteed and it has the risks of disease transmission as well.

The shortcomings of autograft, allograft and DBM justified the development of artificial bone grafts — biomaterials. In the last century, different biomaterials, including metals, polymers, calcium phosphate biomaterials, bioglasses and combinations of thereof, were selected, studied, tested and applied clinically for bone repair alone or with the combination of osteogenic cells or growth factors [3-11].

2. Osteoconductive Biomaterials

As the main inorganic component of bone, calcium phosphate has long been applied in medical use. It was used as a medicine to stimulate bone regeneration in 1890s, but no positive result was obtained until Albee found in 1920s that tricalcium phosphate stimulated bone formation. The real progress of calcium phosphates in medical use took place in 1970-80s when bioglass (CaP containing glass) and hydroxyapatite ceramic (HA) were found to be osteoconductive [12-16]. Thereafter calcium phosphate biomaterials became the most interesting artificial bone grafts.

Osteoconduction has different meanings in different fields. In the clinic, osteoconduction means bone formation towards to the implants from host bone bed. By this meaning, any biomaterial, not only calcium phosphate biomaterials but also porous polymer, can be osteoconductive due to the regeneration ability of bone itself [4]. In biomaterials science, osteoconduction has its special meaning. In addition to the bone formation towards to the implants from the host bone bed (Figure 1 A), osteoconduction in biomaterials science means guided bone formation on material surfaces resulting in bone bonding (Figure 1 B). The latter property is sometimes called "bioactivity", "osteocoalesce" and "osteointegration" [16,17]. In this overview, osteoconduction highlights the guided bone formation on biomaterial surfaces and the chemical bond between newly formed bone and such a biomaterial surface. In practice, the absence of other tissues between the newly formed bone and biomaterial surface is used to identify the biomaterial as osteoconductive (Figure 1 B).

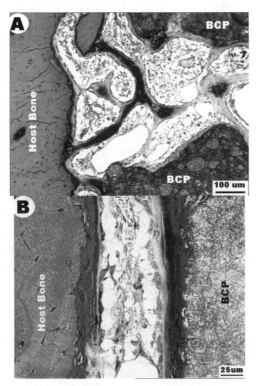

Figure 1. Osteoconduction of calcium phosphate biomaterials. A: Bone ingrowth from the host bone bed into the implant. B: Bonding osteogenesis and bone bond between newly formed bone and the implant surface. (A porous biphasic calcium phosphate ceramic, BCP, 5mm in diameter, was implanted in femoral cortex of dogs for 2 weeks, non-decalcified section, methylene blue and basic fuchsin staining).

After guided bone formation and bone bonding was found on Bioglass and HA, other biomaterials have been reported to have the ability to guide bone formation on their surface and to form a chemical bond to newly formed bone [12-42]. Besides HA ceramics, other calcium phosphate ceramics including biphasic calcium phosphate ceramic (BCP) and tricalcium phosphate ceramic (TCP) have been demonstrated to be osteoconductive. In addition, calcium phosphate cements have been demonstrated to be osteoconductive as well. Calcium phosphate coatings have been used to render metallic surfaces osteoconductive. Also CaP based composites can be made osteoconductive. Actually, even biomaterials that can be easily mineralised *in vivo,* i.e. after implantation, are osteoconductive.

It is becoming common knowledge that osteoconduction is a general property of calcium phosphate biomaterials [12-17]. The physicochemical properties of biomaterials and the peculiarity of osteogenic cells are the key factors causing osteoconduction [17]. Any osteoconductive biomaterial forms a biological apatite layer *in vivo* on its surface. Osteogenic cells then easily attach on the *in vivo* formed apatite surface and form bone on it.

Osteoconductive biomaterials are good bone grafting materials that ensure the continuity of the repaired bone both structurally and mechanically [12-17]. Now, osteoconductive biomaterials have not only been developed in different ways (calcium phosphate ceramics, calcium phosphate cements, calcium phosphate coatings, bioglass, bioglass ceramics, and composites with calcium phosphate) but also in different forms (porous or dense blocks, particles or granules), in order to fulfil the requirements in clinics.

2.1. CALCIUM PHOSPHATE CERAMICS

The more often-used calcium phosphate ceramics are hydroxyapatite ceramic (HA), tricalcium phosphate ceramic (TCP) and the mixture of those two, biphasic calcium phosphate ceramics (BCP with both HA and TCP at different ratio) [14-20]. Though their composition and forms varies, all calcium phosphate ceramics are biocompatible, osteoconductive, non-toxic, antigenically inactive, non-carcinogenic [16-17]. The various calcium phosphate ceramics differ in their dissolution behaviour (Figure 2 A) and therefore their bioresorption rates [14-20]. Hydroxyapatite ceramics (HA), derived from coral or fabricated from synthetic apatite powders, dissolve very slowly, while tricalcium phosphate ceramics (TCP) dissolve faster (Figure 2 A). The dissolution rate of BCP is between HA and TCP (Figure 2 A) and is very much dependent on the TCP/HA ratio: the more TCP, the higher the dissolution rate [14-20].

Besides the higher (physico-chemical) dissolution rate, TCP can be resorbed *in vivo* by osteoclasts or macrophages (Figure 2, B). Although the resorption rate of TCP is affected by both macrostructure and microstructure, TCP ceramic is generally considered as a bioresorbable calcium phosphate ceramic [14-20].

Calcium phosphate ceramics were used as dense blocks, porous blocks, powders or particles, in orthopaedic and dental surgery, also applied as BMP carrier and bone tissue engineering scaffold. Due to their poor mechanical properties, calcium phosphate ceramics are only applied in non-loading bony sites. Moreover, calcium phosphate ceramic blocks are too brittle to be trimmed such that they fill the bone defect perfectly, the consequence being retardation of the bone formation. Ceramic particles or granules fill the bone bed well, but confront the problem of particle migration.

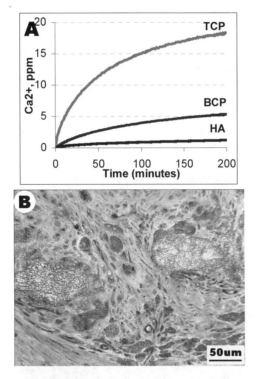

Figure 2. Resorption of calcium phosphate ceramics. A: *in vitro* dissolution of HA, BCP and TCP ceramics (600±10mg ceramic particles, 2-4mm, in 100ml simulated physiological solution, pH=7.3, 37.3°C). B: Bioresorption of TCP ceramic *in vivo* (Porous TCP particles, 2-4mm, implanted in muscle of goats for 6 weeks, non-decalcified section, methylene blue and basic fuchsin staining).

2.2. BIOGLASS AND GLASS CERAMICS

Bioglasses and bioactive glass ceramics were demonstrated to be osteoconductive [12,13,21-23]. Bioglass and glass ceramics are classified as osteoconductive, because of the formation of a CaP rich layer on the surface of bioglass and bioactive glass ceramics. Thus bioglasses and bioactive glass ceramics could be considered as calcium phosphate biomaterials [22,23]. Several kinds of bioglass or bioglass ceramics were developed as granules, particles or bulks and applied clinically alone or as composites with other biomaterials [23].

Bioglass and glass ceramics have similar properties as calcium phosphate ceramics and similar limitations in clinics, but some glass ceramics have a higher mechanical strength than calcium phosphate ceramics [23].

2.3. CALCIUM PHOSPAHTE CEMENT

The limitation of ceramics (either calcium phosphate ceramics or glass ceramics) gave rise to the development of calcium phosphate cement [24-28]. Calcium phosphate cements generally have two components, cement powder and cement solution. Cement powders are normally mixtures of different calcium phosphates. When mixed with cement solution, calcium phosphates hydrolyse and new crystals form. The formed crystals entangle each other, then the cement paste sets and becomes hard. By proper selection of cement compositions (both cement powders and cement solutions), particle size of cement powders and the ratio of cement powder to cement solution, the setting time, mechanical strength, resorption rate of calcium phosphate cement can be adjusted. Several calcium phosphate cements have been invented and they were found to be biocompatible, osteoconductive and to have sufficient strength as a bone filler and a controllable resorption rate [25-28]. A creeping substitution causes resorption of calcium phosphate cement [28]. *In vivo*, calcium phosphate cement is recognized as part of bone, and is resorbed by osteoclasts as in bone remodeling (Figure 3).

Being handled as paste, calcium phosphate cements allow perfect filling of bone defect regardless the shape of the defect, but calcium phosphate cements are still not strong enough for load-bearing bone repair and their resorption *in vivo* is not always predictable.

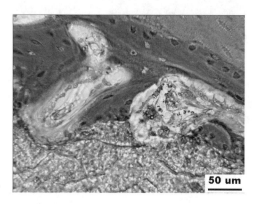

Figure 3. Bioresorption of a calcium phosphate cement (A calcium phosphate cement was implanted in femur of dog for 6 months, non-decalcified section, methylene blue and basic fuchsin staining) [28].

2.4. CALCIUM PHOSPHATE COATINGS

In order to get strong osteoconductive biomaterials for load-bearing bone repair, coating techniques were developed to make non-osteoconductive, but strong, biomaterials osteoconductive [29-37].

The most popular coating technique is plasma spraying [29-32]. With this coating technique, HA, BCP, TCP and even bioglass can be coated onto metal surface to make a metal osteoconductive. Calcium phosphate coated metallic devices including dental roots and hip joints are commercially available for clinical use. However, plasma coating technique is limited to coat the outer surface of metal devices with a simple shape [32]. Thus far, new coating techniques are being developed to make stable calcium phosphate coating onto the surface of biomaterials with complicated structures.

The most recently developed coating technique is the so-called biomimetic coating in which a calcium phosphate apatite layer is formed from aqueous solution at ambient temperature [33,34]. By using this newly developed technique, uniform, stable calcium phosphate coatings can be developed onto the surface of biomaterials regardless the shape, the structure and the chemistry of the biomaterials (not only metals can be coated but also polymers). Due to the fact that biomimetic coatings are made at room temperature, it is possible to incorporate biological factors into calcium phosphate coating in order to improve the biological activity of the coated biomaterials [35,36].

In addition to the *in vitro* biomimetic calcium phosphate coatings to make metals osteoconductive, it is also possible to physicochemically modify metals themselves to accelerate the *in vivo* surface calcification of metals and thereafter make metals osteoconductive [37].

2.5. CALCIUM PHOSPHATE COMPOSITES

In addition to calcium phosphate coatings, osteoconducitve biomaterials can be composed with other biomaterials to make osteoconductive composites. Calcium phosphate ceramics, bioglass or glass ceramics, combined with polymers or collagens as particles, are still osteoconductive. Composites of osteoconductive biomaterials and non-osteoconductive biomaterials are both osteoconductive and better from a mechanical point of view [38-41].

In the future, the introduction of nanotechnology may allow the preparation of osteoconductive composites at nano-scale, and therefore produce osteoconductive composites having mechanical properties comparable to natural bone [42,43].

3. Osteoinductive Biomaterials

Osteoconductive biomaterials are good bone grafting materials, they act as templates for bone formation and form a direct bond with bone. However, osteoconductive biomaterials only passively support bone regeneration but not positively stimulate bone formation. The guided bone formation on osteoconductive biomaterial surface is limited in its distance, and therefore osteoconductive biomaterials alone may not repair large bone defects. For large bone defect repair, bone formation far from the host bone bed should occur by osteoinduction. It was generally thought that biomaterials can only be osteoconductive but not osteoinductive [2-17], however, recent research has demonstrated that osteoinductive biomaterials do exist. For example, when a porous BCP ceramic was implanted in muscle of dogs, ample bone was formed inside the implants after 90 days (Figure 4).

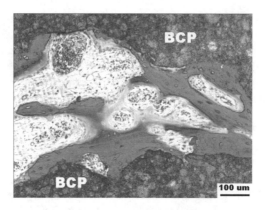

Figure 4. An example of bone formation induced by biomaterials. (A porous BCP ceramic with micropores on macropore surface was implanted in muscle of dog for 3 months, non-decalcified section, methylene blue and basic fuchsin staining)

3.1. OSTEOINDUCTION: DEFINITION.

Osteoinduction is a kind of bone formation that does not start directly from osteogenic cells. It includes two steps, firstly cell differentiation from non-osteogenic cells to osteogenic cells and secondly bone morphogenesis. The most typical example of osteoinduction is bone formation induced by morphogenetic proteins (BMPs) or a BMP-containing matrix [44,45]. Bone formation in soft tissues (muscle or subcutis) where no osteogenic cell exists gives the true indication of osteoinduction and therefore soft tissue implantation is often used to investigate the osteoinductive property of biomaterials. If a biomaterial causes bone formation after implantation in a non-osseous site, the biomaterial is defined as an osteoinductive biomaterial. The bone formation itself is called ectopic bone formation, heterotopic bone formation or heterotopic ossification.

3.2. OSTEOINDUCTION OF CALCIUM PHOSPAHTE BIOMATERIALS

In 1990, for the first time, ectopic bone formation was reported to be induced by calcium phosphate biomaterials. When hydroxyapatite ceramic was implanted subcutaneously in dogs, Yamasaki found bone formation [46]. Subsequently, in 1991, Zhang found bone formation induced by hydroxyapatite ceramic in muscles of dogs and Ripamonti reported bone induction by hydroxyapatite ceramic in Baboons [47,48], and in 1992, Vagervick reported bone induction by hydroxyapatite ceramic in monkeys [49]. Later on, bone formation induced by calcium phosphate ceramic other than hydroxyapatite ceramic was reported in biphasic calcium phosphate ceramic (TCP/HA), α-calcium pyrophosphate ceramic, β-calcium pyrophosphate ceramic, TCP ceramic, calcium phosphate cement, glass ceramic and calcium phosphate-coated metals in dogs [28,46-70]. Not only in baboons, monkeys and dogs, but also in pigs, goats, sheep, rabbits, mice and even human beings [28,46-70] osteoinduction was reported.

3.3. OSTEOINDUCTIVE POTENTIALS

Although numerous calcium phosphate biomaterials have been reported to induce bone formation in non-osseous sites, their osteoinductive potentials were different. Some calcium phosphate biomaterials induced an earlier bone formation and more bone formation than others [46-70]. For example, both HA and BCP induced bone formation in muscles of dogs, but bone formation was found in BCP after 30 days, while bone formation was found in HA not after 30 days but after 45 days (Figure 5). More bone was formed in BCP than in HA at time periods of 45, 60 and 90 days. Osteoinductive potential, which indicates the ability of biomaterials to induce bone formation as characterized by the starting time and amount of the bone induced, could be applied to compare the osteoinductive capacity of biomaterials. Since BCP induced an earlier bone formation and more bone formation than HA, BCP ceramic has a higher osteoinductive potential than HA.

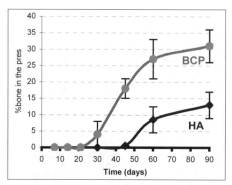

Figure 5. A comparison of osteoinductive potentials of two calcium phosphate ceramics of HA and BCP. (Both ceramics have similar macrostructures and microstructures, were implanted as cylinders, ∅5x6mm, in muscle of dogs for different time periods, the induced bone was measured as the percentage of bone in the pores. Note the starting time of bone formation and the amount of the induced bone in HA and BCP) [70].

3.4. THE MATERIAL FACTORS

The variation of osteoinductive potentials among different calcium phosphate biomaterials suggests that osteoinduction of calcium phosphate biomaterials is material-dependent. Several material factors relevant to osteoinductive potentials have been found.

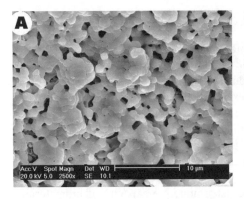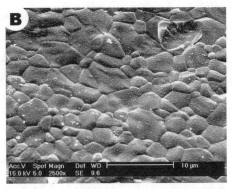

Figure 6. Scanning electronic microscopical oberservation of an osteoinductive HA ceramic (A) and a non-osteoinductive HA ceramic (B). (Note the micropores in osteoinductive HA and the absence of micropores in non-osteoinductive HA) [60].

The most important material factors are the geometrical parameters of biomaterials [58-61,69]. Firstly, a three dimensional environment is needed for osteoinduction to occur. Bone formation induced by biomaterials always occurs inside concave pores of a porous material, or in concave deep rugged surface irregularities of dense materials [58,61]. Secondly, a microstructured surface is needed [58,60,69]. When two porous HA ceramics were implanted in muscles of dogs, HA ceramic with micropores on its macropore surface (Figure 6 A) induced bone formation, while HA without micropores on macropore surface (Figure 6 B) did not induce any bone formation [60]. Another proof of the importance of microstructures in osteoinduction is the influence of sintering temperature on the osteoinductive potentials of calcium phosphate ceramics (Figure 7). When BCP ceramics with the same chemistry were sintered at different temperature of 1100°C, 1200°C and 1300°C, their macroporosity (>50μm) did not change but their microporosity (0.1-10μm) decreased sharply with sintering temperature (Figure 7 A), and so did their osteoinductive potentials (Figure 7 B). When sintered at 1300°C, BCP ceramic had only 1% micropores and did not induce bone formation after 90-day implantation in muscles of dogs, while ceramic sintered at 1100°C had a microporosity of 19% and induced ample bone formation (Figure 7 A and B).

As already mentioned, besides geometrical parameters, the chemistry of biomaterials plays a role in osteoinduction as well. If exhibiting the same macropores and micropores, BCP has a higher osteoinductive potential than HA (Figure 5), obviously related to the higher dissolution rate of BCP than that of HA [58]. But it does not mean that the higher dissolution rate the calcium phosphate biomaterials have, the more osteoinductive they are. Too high dissolution rate results in no bone induction, for example with highly soluble α-TCP ceramic [58,63].

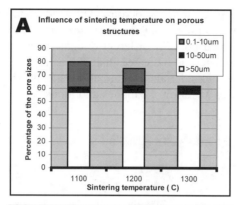

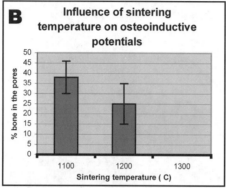

Figure 7. Influence of sintering temperature on mciropoorosity and osteoinductive potentials of BCP ceramics. A: Sharp decrease of microporosity (0.1-10μm) with sintering temperature and no change of macroporosity. B: Decrease of osteoinductive potentials with sintering temperature. (All ceramics were implanted as cylinders, ⌀5x6mm, in muscle of dogs for 3 months, osteoinductive potentials were measured as the percentage of induced bone in the pores).

3.5. ANIMAL-DEPENDENCE

At first, bone formation induced by calcium phosphate biomaterials was found in specific animals such as baboons, monkeys, dogs but not in general experimental animal models such as rabbits, rats and mice, and therefore, it was suggested that osteoinduction of calcium phosphate biomaterials is animal-dependent and might be limited to some specific animals [46-54]. The improvement of calcium phosphate biomaterials with regard to their osteoinductive potentials has made osteoinduction in general experimental animal models possible by calcium phosphate biomaterials. Both HA and BCP ceramics induce bone formation in muscles of dogs, but only BCP that has a higher osteoinductive potential than HA induced bone formation in soft tissues of rabbits and mice [66]. Compared to the bone induced by BCP in dogs, bone induced by BCP in rabbits and mice is later and less [66]. Bone induction has so far been found in many animals including baboons [47, 54, 61], monkeys [49], pigs [55], dogs [46,48,50-52, 54-56,60,62-66,70], goats [59,69], sheep [67], rabbits [57,66,68], mice and even in human beings [53]. It could be concluded that osteoinduction of calcium phosphate biomaterials is a general phenomenon in mammals provided that calcium phosphate biomaterials have certain osteoinductive generating properties. Although bone could be induced by calcium phosphate biomaterials in any mammal, the variation of bone incidence and the amount of the formed bone in different animal species induced by the same calcium phosphate biomaterials indicated that in quantitative terms osteoinduction of calcium phosphate biomaterials is animal-dependent [52,54,66]. In general, osteoinduction decreases as follows: baboons and monkeys > pigs and dogs > goats and sheep > rabbits and mice.

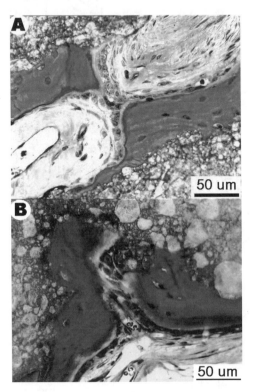

Figure 8. Bone formation induced by a BCP ceramic in rabbits (A) and mice (B). (Ceramic was implanted as cylinders, ∅5x6mm in muscle of rabbit, ∅4x5mm under skin of mice, for 3 months, non decalcified sections, methylene blue and basic fuchsin staining).

3.6. TIME COURSE OF MATERIAL-INDUCED BONE FORMATION

Several steps are followed in bone formation induced by osteoinductive biomaterials. Among them are 1) attachment of mesenchymal cells on material surface, 2) proliferation and differentiation of mesenchymal cells, 3) bone matrix formation by induced osteogenic cells, 4) mineralization of bone matrix and 5) bone remodeling to form mature bone (Figure 9). The time for different steps to happen varies with the osteoinductive potentials of calcium phosphate biomaterials. In BCP ceramic having a higher osteoinductive potential, mesenchymal cells have attached on BCP surface two weeks after implantation, cells have proliferated and differentiated on BCP surface 3 weeks after implantation (Figure 9 A), and bone matrix has been formed within 30 days after implantation (Figure 9 B). Mineralized bone has been formed in 45days after implantation (Figure 9 C) and bone remodeling has started within 60 days (Figure 9 D). In HA ceramic which has a lower osteoinductive potential than BCP, the processes of bone induction postponed 2 weeks (Figure 5). When a non-osteoinductive biomaterial is implanted in soft tissue, the processes of osteoinduction will never happen.

Although the starting time for bone formation induced by different biomaterials varies, similar steps are always followed. Bone formation always starts directly as bone on calcium phosphate biomaterials surface, chondrocytes and cartilage formation have never been found. Bone naturally occurs in development either by direct replacement of primitive connective tissues (intramembranous ossification) or by replacement of pre-existing cartilage (endochondral ossification). Since cartilage formation has never been found in bone formation induced by osteoinductive calcium phosphate biomaterials, it is supposed that bone formation induced by osteoinductive biomaterials resembles the intramembranous ossification in development [47,54,56]. However, it may be noted that bone formation direct on the resorbed bone bed during bone remodeling (the secondary bone formation) starts directly as bone as well.

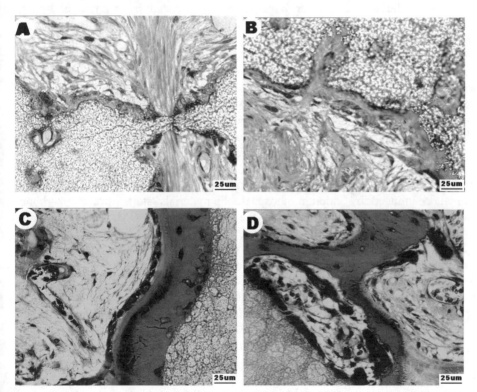

Figure 9. Historical events in osteoinduction. A: Cell attachment and differentiation on an osteoinductive biomaterial surface (21days). B: Osteoid formed on osteoinductive biomaterial surface (30days). C: Mineralization of bone matrix on osteoinductive biomaterial surface (45days). D: Bone remodeling by osteoclasts (60days). (BCP was implanted in muscle of dogs, non-decalcified sections, methylene blue and basic fuchsin staining).

In the secondary bone formation, osteogenic precursor cells attach on the resorbed bone bed, proliferate, differentiate and form new bone. Bone formation induced by osteoinductive calcium phosphate biomaterials has the similar sequences as the secondary bone formation does. Before and when bone formation induced by biomaterials starts, osteogenic precursor cells (mesenchymal cells) attach on osteoinductive biomaterial surface, proliferate and differentiate to osteogenic cells which then form bone. It is thus more likely that bone formation induced by osteoinductive calcium phosphate biomaterials is a secondary bone formation [70].

3.7. THE MECHANISM OF MATERIAL-INDUCED BONE FORMATION

Osteoinductive calcium phosphate biomaterials are becoming interesting to scientists, however, most studies are limited to getting more evidence of osteoinduction of calcium phosphate biomaterials or how to improve osteoinductive calcium phosphate biomaterials. The mechanism of bone formation induced by calcium phosphate biomaterials is not addressed yet and the current explanations of the mechanism are hypothetical rather than conclusive. Bone morphogenetic proteins (BMPs), which is the reason of most osteoinduction cases, is usually supposed to be the reason for bone formation induced by calcium phosphate biomaterials [54,61], due to the fact that calcium phosphate biomaterials have a strong affinity to BMPs [44,45] naturally present in body fluids. It is impossible to exclude BMPs from bone formation induced by calcium phosphate biomaterials, because where is bone formation, there is BMPs. It follows then that the causative role of BMPs in bone formation induced by calcium phosphate biomaterials can not be conclusive unless the presence of BMPs in calcium phosphate biomaterials before cell differentiation (from non-osteogenic cells to osteogenic cells) has been obtained. Even if it were clearly been shown that BMPs are present in osteoinductive calcium phosphate biomaterials before cell differentiation, BMPs could not be the only reason for bone formation induced by calcium phosphate biomaterials, since bone formation induced by BMPs starts as cartilage while bone formation induced by calcium phosphate biomaterials starts directly as bone. Another argument to cast doubt on the crucial role of BMP's in explaining osteoinductive property of biomaterials is that bone formation can be easily induced by BMP's in small animals but difficult so in large animals, while bone induction by calcium phosphate biomaterials is just easier in large animals.

As discussed above, bone formation induced by calcium phosphate biomaterials is more likely the secondary bone formation as occurring in bone remodeling. Through an *in vivo* calcification, a cementline was formed on the resorbed bone bed and the secondary bone formation starts [17]. The *in vivo* surface calcification may play important role in the secondary bone formation and in bone formation induced by calcium phosphate biomaterials as well. Actually, the role of in vivo calcification in bone formation induced by biomaterials has been shown. In 1969, Simpson *et al* found bone formation induced by a polymer material implanted in soft tissue of pigs, in which they clearly observed that bone formation started after the polymer surface was calcified [71]. More recently, Kokubo *et al* have got bone formation induced by a porous metal having super ability to form CaP layer on its surface [72]. Although more evidence is needed, it appears that the *in vivo* calcification of biomaterials surface may be the key reason for bone formation induced by calcium phosphate biomaterials.

Besides BMP's and *in vivo* surface calcification, other factors may be involved in bone formation induced by calcium phosphate biomaterials, including growth factors other than BMPs, the surface energy (zeta-potentials), low oxygen, asymmetric cell division which may occur on a rough surface and cause cells to differentiate into osteogenic cells.

3.8. THE SIGNIFICANCE OF OSTEOINDUCTIVE BIOMATERIALS

Osteoinductive calcium phosphate biomaterials support osteogenic cells to form bone [73]. When *in vitro* expanded goat bone marrow stromal cells were cultured on a non-osteoinductive calcium phosphate ceramic (HA) and an osteoinductive calcium phosphate ceramic (BCP) for 1 day and thereafter autologously implanted in muscles of goats for 12 weeks, cells on osteoinductive BCP ceramic gave ample bone formation (Figure 10 A), while cells on non-osteoinductive HA ceramic did not give any bone formation (Figure 10 B).

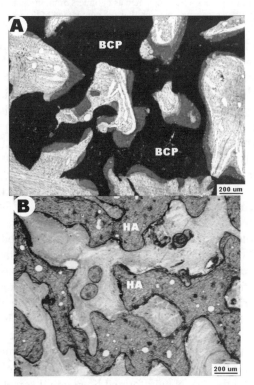

Figure 10. Histology of tissue-engineered bone after in vivo implantation. A: Bone formation of tissue-engineered bone with osteoinductive BCP. B: Absence of bone formation in tissue-engineered bone with non-osteoinductive HA. (In vitro expanded goat bone marrow stromal cells were cultured on BCP and HA for 1day and autologously implanted in muscle of goats for 12 weeks, non-decalcified sections, metheylene blue and basic fuchsin staining) [73].

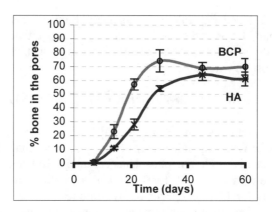

Figure 11. Bone repair with osteoinductive biomaterials. (BCP and HA, having different osteoinductive potentials as indicated in Figure 5, were implanted as cylinders 5mm in diameter, in femur of dogs for different time periods, bone repair was measured as the percentage of the formed bone in pores) [74].

The higher osteoinductive potential the implant materials have, the faster bone repair [74]. Osteoinductive calcium phosphate biomaterials in the one hand provide a good environment for osteogenic cells to form bone and on the other hand induce non-osteogenic cells to differentiate into osteogenic cells that then form bone. Theoretically, osteoinductive calcium phosphate biomaterials give faster bone formation in bone repair. Fast bone formation in theory happened in practice as shown in a comparison of two calcium phosphate ceramics (BCP and HA). As discussed previously, BCP ceramic has a higher osteoinductive potential than HA when exhibiting similar porous structures (both macropores and micropores). When ceramic cylinders (5mm in diameter) of both BCP and HA were implanted intracortically in femurs of dogs for different time periods of 7days, 14day, 21day, 30days, 45days and 60days, more bone was formed in BCP than HA at day 14, day 21 and day 30 (Figure 11). The bone defects have been repaired by BCP in 30 days after implantation because the formed bone remained stable after 30days, while the bone defects were repaired two weeks later with HA than with BCP.

In addition to the fast bone repair in small bone defects, osteoinductive calcium phosphate biomaterials are expected to repair critical bone defects, since bone formation occur far from the host bone bed by osteoinduction in large bone defects. When an osteoinductive BCP ceramic was used to repair a well-established critical bone defect (17mm in diameter) in iliac wing of goats, after 12 weeks bone formation has already forwarded to 6.3±1.7mm far from the host bone bed (74% of the total length of the defect). Newly formed bone could sometimes even be found throughout the implant (100% of the total length) (Figure 12). Compared to the bone repair with allograft and autograft in the same model at the same time period, bone formation in osteoinductive BCP is much better than allograft and equal to autograft [75].

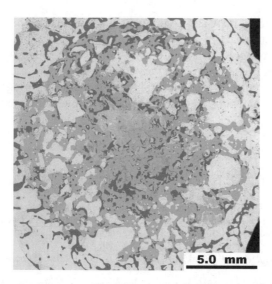

Figure 12. Repair of a critical defect in iliac wing of goat. (Osteoinductive BCP ceramic discs, Ø17x6mm, were implanted in iliac wing of goats for 12 week, a pseudocolored image, Red: bone, Green: BCP) [75].

4. The Future Challenges

It is obvious that calcium phosphate biomaterials can be made osteoinductive through physicochemical modification besides by additional introduction of growth factors or osteogenic cells. It appears that osteoinductive calcium phosphate biomaterials are good biomaterials for bone repair. The osteoinductive potentials of calcium phosphate biomaterials are material-dependent. Some material factors affecting osteoinductive potentials have been identified, but not fully understood. The full understanding of material factors relevant to bone formation induced by calcium phosphate biomaterials may result in calcium phosphate biomaterials with higher and higher osteoinductive potentials which can induce bone formation as early as possible and induce as much bone formation as possible, and therefore repair bone as fast as possible and as large as possible.

Apart from osteoconductive property and osteoinductive property, bone grafts require other properties. An ideal bone graft should be osteoconductive, osteoinductive, resorbable, easy to shape and have good mechanical properties. New techniques are necessary to implement calcium phosphate biomaterials according to the clinical requirements.

How far away osteoinductive calcium phosphate biomaterials can go in bone repair is not known as yet. However, even though osteoinductive potentials of calcium phosphate biomaterials alone are still not effective enough for bone repair, osteoinductive biomaterials could be used as bone tissue engineering scaffold, growth factor carrier, gene carrier or extender of autograft.

From no bone to bone, bone formation induced by calcium phosphate biomaterials serves a good model to study bone biology and bone physiology. The mechanism of bone formation by osteoinductive biomaterials is not clear as yet, however, the more understanding of such mechanism in the future may, in the one hand help us

54

understand more about bone and bone repair, and on the other hand help us improve osteoinductive biomaterials.

At last but not least, graft is only a part of bone repair, other things including the instrumentation and fixation are also very important for bone regeneration and bone repair [3-11]. In other words, a bridge is needed to bring osteoconductive, osteoinductive calcium phosphate biomaterials to clinics for bone repair.

References

1. Bloom, W. and Fawcettt, D.W. (1986) *A textbook of histology*, W. B. Saunders, Philadelphia, pp. 199-238.
2. Rosenberg, A. (1999) Bones, joints, and soft tissue tumors, in R.S. Cotran, V. Kumar and T. Collins (eds) *Robins pathological basis of disease*, W. B. Saunders, Philadelphia, pp. 1215-1268.
3. Damien, C.J. and Parsons, J.R. (1991) Bone graft and bone graft substitutes: A review of current technology and application, *J.Appl.Biomat.* **2**, 187-208.
4. Hollinger, J.O., Brekke, J., Gruskin, E, and Lee, D. (1996) Role of bone substitutes, *Clin.Orthop.* **324**, 55-65.
5. Perry, C.R. (1999) Bone repair techniques, bone graft, and bone graft substitutes, *Clin.Orthop.* **360**, 71-86.
6. Betz, R.R. (2002) Limitations of autograft and allograft: new synthetic solutions, *Orthopedics* **25**, s561-570.
7. Parikh, S.N. (2002) Bone graft substitutes: past, present, future, *J. Postgrad. Med.* **48**,142-8.
8. McAuliffe, J.A. (2003) Bone graft substitutes, *J. Hand Ther.* **16**,80-187.
9. Costantino, P.D., Hiltzik, D., Govindaraj, S. and Moche, J. (2002) Bone healing and bone substitutes, *Facial Plast. Surg.* **18**, 13-26.
10. Sammarco, V.J. and Chang, L. (2002) Modern issues in bone graft substitutes and advances in bone tissue technology, *Foot Ankle Clin.* **7**, 19-41.
11. Bucholz, R.W,. (2002) Nonallograft osteoconductive bone graft substitutes, *Clin. Orthop.* **395**, 44-52
12. Hench, L.L. (1980) Biomaterials, *Science* **208**, 826-831.
13. Jarcho, M. (1981) Calcium phosphate ceramics as hard tissue prosthetics, *Clin.Orthop.* **157**, 259-278.
14. Hench, L.L. and Wilson, J. (1984) Surface-active biomaterials, *Science* **226**, 630-635.
15. de Groot K. (1984) Calcium phosphate ceramics: their current status, in J.W. Boretos, and M. Eden (eds), *Contemporary Biomaterials*. Noyes Publications, USA, pp. 477-492.
16. Daculsi, G. and Passuti, N. (1989) Bioactive ceramics, fundamental properties and clinical applications: the osteo-coalescence process, *Bioceramics* **2**, 3-10.
17. Osborn, J.F. (1991) The biological profile of hydroxyapatite ceramic with respect to the cellular dynamics of animal and human soft tissue and mineralized tissue under unloaded and loaded conditions, in M.A. Barbosa (eds), *Biomaterials Degradation*, Elsevier, Amsterdam, pp. 185-225.
18. Daculsi, G. (1998) Biphasic calcium phosphate concept applied to artificial bone, implant coating and injectable bone substitute, *Biomaterials* **19**, 1473-1478.
19. Szpalski, M., and Gunzburg, R. (2002) Applications of calcium phosphate-based cancellous bone void fillers in trauma surgery, *Orthopedics* **25**, s601-609.
20. Metseger, D.S. and Driskell, T.D. (1982) Tricalcium phosphate ceramic, a resorbable bone implant: Review and current status, *J. Am. Dent.Assoc.* **105**, 1035-1044.
21. Nakamura, T., Yamamura, T., Higashi, S., Kokubo, T. and Itoo, S. (1985) A new glass-ceramic for bone replacement: Evaluation of its bonding to bone tissue, *J. Biomed. Mater. Res.* **19**, 685-698.
22. El-Ghannam, A., Ducheyne, P. and Shapiro, I.M. (1997) Formation of surface reaction products on bioactive glass and their effects on the expression of osteoblastic phenotype and the deposition of mineralized extracellular matrix, *Biomaterials* **18**, 295-303.
23. Kokubo, T., Kim, H.M. and Kawashita, M. (2003) Novel bioactive materials with different mechanical properties, *Biomaterials* **24**, 2161-2175.
24. Ginebra, M.P., Fernandez, E., Boltong, M.G., Planell, J.A., Bermudez, O. and Driessens, F.C.M. (1994) Compliance of a calcium phosphate cement with some short-term clinical requirement, *Bioceramics* **6**, 273-278.

25.   Miyamoto, Y., Ishikawa, K., Fukao, H., Sawada, M., Nagayama, M., Kon, M. and Asaoka, K. (1995) In vivo setting behaviour of fast-setting calcium phosphate cement, *Biomaterials* **16**, 855-860.

26.   Miyamoto, Y., Ishikawa, K., Takechi, M., Yuasa, M., Kon, M., Nagayama, M. and Asaoka, K. (1996) Non-decay type fast-setting calcium phosphate cement: setting behavior in calf serum and its tissue response, *Biomaterials* **17**, 1429-1435.

27.   Constantz, B.R., Barr, B.M., Ison, I.C., Fulmer, M.T., Baker, J., McKinney, L., Goodman, S.B., Gunasekaren, S., Delaney, D.C., Ross, J. and Poser, R.D. (1998) Histological, chemical, and crystallographic analysis of four calcium phosphate cements in different rabbit osseous sites, *J. Biomed. Mater. Res.* **43**, 451-461.

28.   Yuan, H., Li, Y., de Bruijn, J.D., de Groot, K. and Zhang, X. (2000) Tissue responses of calcium phosphate bone cement: A study in dogs, *Biomaterials* **21**, 1283-1290.

29.   de Groot, K., Geesink, R., Klein, CPAT and Serekian, P. (1987) Plasma sprayed coatings of hydroxyapatite, *J. Biomed. Mater. Res.* **21**,1375-1381.

30.   Cook, S.D., Thomas, K.A., Kay, J.F. and Jarcho, M. (1988) Hydroxyapatite-coated porous Titanium for use as an orthopedic biologic attachment system, *Clin.Orthop.* **230**, 303-311.

31.   Vercaigne, S., Wolke, J.G.C., Naert, I. And Jansen, J.A. (1998) Bone healing capacity of titanium plasma-sprayed and hydroxyapatite-coated oral implants, *Clin. Oral implants Res.* **9**, 261-271.

32.   Lacefield, W.R. (1998) Current status of ceramic coatings for dental implants, *Implant Dent.* **7**, 315-322.

33.   Abe, Y., Kokubo, T. and Yamamura, T. (1990) Apatite coating on ceramics, metals and polymers utilizing a biological process, *J.Mater.Sci: Mater.Med* **1**, 233-238.

34.   Leitao, E., Barbosa, M.A. and de Groot, K. (1995) In vitro calcification of orthopaedic implant materials, *J. Mater. Sci: Mater. Med* **5**, 849-852.

35.   Wen, H.B., de Wijn, J.R,. van Blitterswijk, C.A. and de Groot, K. (1999) Incorporation of bovine serum albumin in calcium phosphate coating on titanium, *J. Biomed. Mater. Res.* **46**, 245-252.

36.   Liu, Y., Hunziker, E.B., Layrolle, P. and de Groot, K. (2002) Introduction of ectopic bone formation by BMP-2 incorporated into calcium phosphate coatings of Titanium-Alloy implants, *Bioceramics* **15**, 667-670.

37.   Fujibayashi, S., Nakamura, T., Nishiguchi, S., Tamura, J., Uchida, M. and Kim, H.M. (2001) Kokubo T, Bioactive titanium: effect of sodium removal on the bone-bonding ability of bioactive titanium prepared by alkali and heat treatment, *J. Biomed. Mater. Res.* **56**,562-570.

38.   Zardiackas, L.D., Teasdall, R.D., Black, R.J., Jones, J.S., St.John, R., Dillion, L.D. and Hughes, J.L. (1994) Torsional properties of healed canine diaphyseal defects grafted with a fibrillar collagen and hydroxyapatite/tricalcium phosphate composite, *J. Appli. Biomater.* **5**, 277-283.

39.   John, R.K., Zardiackas, L.D., Terry, R.C., Teasdall, R.D. and Cooke, S.E. (1995) Histological and electron microscopic analysis of tissue response to synthetic composite bone graft in the canine, *J. Appl. Biomater.* **6**, 89-97.

40.   Lawson, A.C. and Czernuszka, J.T. (1998) Collagen--calcium phosphate composite, *Proc. Inst. Mech. Eng.* **212**, 413-425.

41.   Muzzarelli, C. and Muzzarelli, R.A. (2002) Natural and artificial chitosan-inorganic composites, *J. Inorg. Biochem.* **92**, 89-94.

42.   Roy, I., Mitra, S., Maitra, A. and Mozumdar, S. (2003) Calcium phosphate nanoparticles as novel non-viral vectors for targeted gene delivery, *Int. J. Pharm.* **250**, 25-33.

43.   Forster, S. and Plantenberg, T. (2002) From self-organizing polymers to nanohybrid and biomaterials, *Angew Chem. Int. Ed. Engl.* **41**, 689-714.

44.   Urist, M.R. (1965) Bone: formation by autoinduction, *Science* **150**, 893-899.

45.   Urist, M.R., Huo, Y.K. and Brownell, A.G. (1984) Purification of bovine bone morphogenetic protein by hydroxyapatite chromatography, *Proc.Natl.Acad.Sci.USA* **81**, 371-375.

46.   Yamasaki, H. (1990) Heterotopic bone formation around porous hydroxyapatite ceramics in the subcutis of dogs, *Jpan. J. Oral Biol.* **32**, 190-192.

47.   Ripamonti, U. (1991) The morphogenesis of bone in replicas of porous hydroxyapatite obtained from conversion of calcium carbonate exoskeletons of coral, *J. Bone & Joint Surg.* **73A**, 692-703.

48.   Zhang, X. (1991) A study of porous block HA ceramics and its osteogenesis, in A. Ravaglioli and A. Krajewski (eds), *Bioceramics and the Human Body*, Elsevier Science, Amsterdam, pp. 408-415.

49.   Vargervik K. (1992) Critical sites for new bone formation, In M.B. Habal and A.H. Reddi (eds), *Bone grafts & bone substitutes*, W.B. Saunders, Philadelphia, pp.112-120.

50.   Yamasaki, H. and Saki, H. (1992) Osteogenic response to porous hydroxyapatite ceramics under the skin of dogs, *Biomaterials* **13**, 308-312.

56

51. Toth, J.M., Lynch, K.L. and Hackbarth, D.A.(1993) Ceramic-induced osteogenesis following subcutaneous implantation of calcium phosphates, *Bioceramics* 6, 9-13.
52. Klein, CPAT, de Groot, K., Chen, W., Li, Y. and Zhang, X. (1994) Osseous substance formation induced in porous calcium phosphate ceramics in soft tissues, *Biomaterials* 15, 31-34.
53. Green, J.P., Wojno, T.H., Wilson, M.W. and Grossniklaus, H.E. (1995) Bone formation in hydroxyapatite orbital implants, *Am. J. Ophthalmol.* 120, 681-682.
54. Ripamonti, U. (1996) Osteoinduction in porous hydroxyapatite implanted in heterotopic sites of different animal models, *Biomaterials* 17, 31-35.
55. Yang, Z., Yuan, H., Tong, W., Zou, P., Chen, W. and Zhang, X. (1996) Osteogenesis in extraskeletally implanted porous calcium phosphate ceramics:variability among different kinds of animals, *Biomaterials* 17, 2131-2137.
56. Yang, Z., Yuan, H., Zou, P., Tong, W., Qu, S. and Zhang, X. (1997) Osteogenic responses to extraskeletally implanted synthetic porous calcium phosphate ceramics: an early stage histomorphological study in dogs. *J. Mater. Sci: Mater. Med.* 8, 697-701.
57. Sires, B.S., Holds, J.B., Kincaid, M.C. and Reddi, A.H. (1997) Osteogenin-enhanced bone-specific differentiation in hydroxyapatite orbital implants, *Ophthal. Plast. Reconstr. Surg.* 13, 244-251
58. Yuan, H., Yang, Z., Li, Y., Zhang, X., de Bruijn, J.D. and de Groot, K. (1998) Osteoinduction by calcium phosphate biomaterials, *J. Mater. Sci: Mater. Med.* 9, 723-726.
59. de Bruijn, J.D., Dalmeijer, R. and de Groot, K. (1999) Osteoinduction by microstructured calcium phosphates, Transaction of 25th Annual meeting of Society for Biomaterials, RI, USA, p235.
60. Yuan, H., Kurashina, K., de Bruijn, J.D., Li, Y., de Groot, K. and Zhang, X. (1999) A preliminary study on osteoinduction of two kinds of calcium phosphate ceramics, *Biomaterials* 20, 1799-1806.
61. Ripamonti, U., Crooks, J. and Kirkbride, A.N. (1999) Sintered porous hydroxyapatite with intrinsic osteoinductive activity: geometric induction of bone formation, *South African Journal of Science* 95, 335-343.
62. de Bruijn, J.D., Yuan, H., Dekker, R. and van Blitterswijk, C.A. (2000) Osteoinduction by biomimetic calcium phosphate coatings and their potential use as tissue engineering scaffolds. in J. E. Davies (eds), *Bone engineering*, em squared incorporated, Toronto, pp.421-431.
63. Yuan. H., de Bruijn, J.D., Li, Y., Feng, J., Yang, Z., de Groot, K. and Zhang, X. (2001) Bone formation induced by calcium phosphate ceramics in soft tissue of dogs: A comparative study between α-TCP and β-TCP, *J. Mater. Sci: Mater. Med.* 12, 7-13.
64. Yuan, H., Yang, Z., de Bruijn, J.D., de Groot, K. and Zhang, X. (2001) Material-dependent bone induction by calcium phosphate ceramics: A 2.5-year study in dog, *Biomaterials* 22, 2617-2623.
65. Yuan, H., de Bruijn, J.D., Zhang, X., van Blitterswijk, C.A. and de Groot, K. (2001) Bone induction by porous glass ceramic made from Bioglass (45S5), *J. Appl. Biomat.* 58, 270-276.
66. Yuan, H., de Bruijn, J.D., van Blitterswijk, C.A. and de Groot, K. (2001) Bone induction by a calcium phosphate ceramic in rabbits. Transaction of 27th Annual meeting of society for biomaterials, Minnesota, USA, pp.142.
67. Gosain, A.K., Song, L.,Riordan, P., Amarante, M.T., Nagy, P.G., Wilson, C.R., Toth, J.M., and Ricci, L. (2002) A 1-year study of osteoinduction in hydroxyapatite-derived biomaterials in an adult sheep model: part I, *Plastic and Reconstructuve Surgery* 109, 619-630.
68. Kurashina, K., Kurita, H., Wu, Q., Ohtsuka, A., and Kobayashi, H. (2002) Ectopic ostepgenesis with biphasic ceramics of hydroxyapatite and tricalcium phosphate in rabbits, *Biomaterials* 23, 407-412.
69. Yuan, H., van den Doel, M., van Blitterswijk, C.A., de Groot, K. And de Bruijn, J.D. (2002) A comparison of bone induction by two kinds of calcium phosphate ceramics in goats, *J. Mater. Sci: Mater. Med.* 13,1271-1275.
70. Yuan, H., de Bruijn, J.D., van Blitterswijk, C.A. and de Groot, K. (2002) Time course of bone induction by an osteoinductive biomaterial, Transaction of 28th Annual meeting of Society for Biomaterials, Tampa, Florida, USA, pp. 706.
71. Winter, G.D. and Simpson, B.J. (1969) Heterotopic bone formed in a synthetic sponge in the skin of young pigs, *Nature* 223, 88-90
72. Fujiibayashi, S., Neo, M., Kim, H.M., Kokubo, T., and Nakamura, T. (2003) Osteoinduction of porous bioactive titanium metal, *Biomaterials*, (in press).
73. Yuan, H., Van den Doel, M., van Blitterswijk, C. A., de Groot, K. and de Bruijn, J.D. (2002) A comparison of two kinds of calcium phosphate ceramics as bone tissue engineering scaffold in goats, Transaction of 17th meeting of European Society for Biomaterials, Barcelona, Spain, T159.

74. Yuan, H., de Bruijn, J.D., van Blitterswijk, C.A. and de Groot, K. (2002) Osteoinductive biomaterials and bone repairs, Transaction of 17th meeting of European Society for Biomaterials, Barcelona, Spain, Barcelona, Spain, pp. 156.

75. Yuan, H., Kruyt, M., van den Doel, M., van Blitterswijk, C.A., de Groot, K. and de Buijn, J.D. (2004) Repair of a critical size bone defect in goat with an osteoinductive calcium phosphate ceramic, The 7th World Biomaterials Conference, Sydney, Australia. Submitted.

NANOSTRUCTURAL CONTROL OF IMPLANTABLE XEROGELS FOR THE CONTROLLED RELEASE OF BIOMOLECULES

SHULA RADIN AND PAUL DUCHEYNE
Department of Bioengineering
Center for Bioactive Materials and Tissue Engineering
University of Pennsylvania

1. Introduction

Controlled release systems are designed to deliver controlled amounts of therapeutic agents to specific target sites over extended duration of time. The local delivery eliminates the risks of side effects associated with oral or parenteral therapies such as systemic toxicity. It also improves the efficacy of the treatment by achieving higher drug concentrations at the target site than are possible with systemic administration. In order to avoid the need for a second operation, resorbable and biocompatible materials are very desirable for topical applications during surgery. In response to the need for such carriers, biodegradable polymers have been proposed [1-6]. However, it has been reported that the degradation of polymers, which is the mechanism controlling the release of bioactive molecules, can cause an inflammatory response that interferes with the intended therapy [6-9]. Demineralized bone matrix, synthetic bioactive ceramics and glass-ceramics have also been considered as implantable controlled release materials [10-16]. These materials are biocompatible and can enhance bone healing. However, these macroporous materials usually demonstrate a poorly controlled "burst" release profile; that is, the release is largely terminated shortly after implantation. For effective healing, however, it is desirable to release an initial large concentration of drugs post-operatively, followed by a steady long-term release.

Sol-gel processed materials represent an alternative to the materials previously used for the controlled release of biologically active agents [17]. Room-temperature processed sol-gel derived silicas (also called xerogels) have been explored for various biomedical applications. Base-catalyzed or two-step acid-base-catalyzed xerogels have been used for the encapsulation of enzymes, cells and living tissues [18-20]. Acid-catalyzed xerogels have been studied as controlled release materials [21-28]. A simple room-temperature process can be used for the incorporation of various biological molecules into silica xerogels. This room-temperature process provides easily reproducible xerogel properties. Large quantities of biological agents can be added and uniformly distributed in the liquid sol. After gelation,

R.L. Reis and S. Weiner (eds.),
Learning from Nature How to Design New Implantable Biomaterials, 59-74.
© 2004 *Kluwer Academic Publishers. Printed in the Netherlands.*

condensation and drying, the agents become encapsulated in a glassy solid. These materials are resorbable, highly porous and nanostructured (pore size from 1 to 5 nm). Their structural properties can be extensively controlled by altering sol-gel processing parameters (such as selection of silica precursors, the use of additional oxides, catalysts, solvents, pH of sol, water/alkoxide molar ratio, temperature and conditions of condensation and drying) [17].

In our laboratory, we have studied room temperature processed silica xerogels for controlled release of various biological agents such as drugs, proteins and growth factors. [22-25] In vitro, we studied the effect of synthesis parameters on the structure and the release properties of xerogels. In vivo, we assessed silica xerogels as resorbable and biocompatible materials [26]. In this paper, we summarize these in vitro and in vivo studies of silica xerogel controlled released materials.

2. Synthesis of Silica Xerogels with Incorporation of Biologically Active Molecules – Methods

The chemical reactions that take place during the synthesis of silica gels include hydrolysis and condensation reactions. The hydrolysis reaction, which can be either acid or base catalyzed, replaces alkoxide groups with hydroxyl groups. Siloxane bonds (Si-O-Si) are formed during subsequent condensation. Alcohol and water are byproducts of the condensation reaction and evaporate during drying. Theoretically, the overall reaction is as follows [17]:

$$n \, Si(OR)4 + 2n \, H2 \, O \rightarrow n \, SiO2 + 4n \, ROH$$

However, in reality, the completion of the reaction and the chemical composition of the resulting product depend on the excess of water above the stoichiometric H_2O/Si ratio of 2. A number of other sol-gel processing parameters (such as pH of the sol, type and concentration of solvents, temperature, aging and drying schedules, etc.) can also affect the composition, structure, and properties of the resulting product.

In our studies, various biological molecules having different size were used for incorporation into silica xerogels. These included drugs such as naltrexone (0.4 kDa) and the antibiotic vancomycin (3 kDa) and macromolecules such as trypsin inhibitor (20 kDa) and TGF-β (22 kDa) [22-25].

All xerogels were synthesized by a room temperature process using silica precursors such as tetramethoxysilane (TMOS) or tetraethoxysilane (TEOS). In general, the procedure was as follows: single-step acid-catalyzed hydrolysis (pH 2.0–3.0) with excess of water over the stoichiometric H_2O/Si ratio was used. TMOS (or TEOS), deionized (DI) water and acid were mixed in a glass beaker and stirred using a magnetic stirrer to obtain a homogenous sol. HCl or glacial acetic acids were used as catalysts for the hydrolysis of TMOS and TEOS, respectively. Then molecules of interest were then dissolved in DI and added to the sol. Upon mixing, the sol was cast into cylindrical polystyrene vials. The vials were sealed and the sol samples were allowed to gel and age. Subsequently, the vials were opened, and the gels were allowed to dry until the gel weight became constant.

Using the basic single-step acid catalyzed process, a number of processing parameters was varied to determine the effects on the nanostructure and release properties of the xerogels.

These included variations in pH, solvents, water/alkoxide molar ratio, and aging and drying schedule. Modifications of the sol-gel process included the use of organic or organomodified alkoxysilane. Parametrical changes also included variations in concentrations of incorporated bioactive molecules.

2.1. CHARACTERIZATION OF XEROGEL NANOSTRUCTURE

Fourier Transform Infrared (FTIR) spectrometry (5 DXC, Nicolet, Madison, WI) and gas (nitrogen) sorption analysis (Autosorb 1, Quantachrome) were used for the xerogel characterization. FTIR analysis was performed using the diffuse reflectance mode. Prior to obtaining nitrogen adsorption-desorption isotherms the samples were outgassed at 35^0C for 48 hours. Multi-point BET [29] was used to determine the surface area (SA), pore volume (PV) and mean pore radius (PR).

FTIR spectra of xerogels (not shown) were typical for acid-catalyzed silica xerogels [17]. The FTIR analysis showed that the xerogels were well hydrolyzed and polymerized [24]. The incorporation of bioactive molecules did not produce any discernible effect on the FTIR spectra of xerogels [24].

Nitrogen adsorption-desorption isotherms of all acid-catalyzed silica xerogels studied were typical for a Type I isotherm (BDDT classification [30]) [24]. This isotherm type is characteristic of a microporous solid, i.e. a solid with a large number of pores having radii equal to or below 1.5 nm. Variation of the processing parameters (which included the use of methanol as a solvent, pH (from 1 to 5), water/alkoxide ratio (from 4 to 10), aging time (from 1 to 14 days), and drying schedule)) did not affect the isotherm type. The isotherms of xerogels containing bioactive molecules such as naltrexone, vancomycin and trypsin inhibitor were also characteristic of a microporous solid.

Physical properties of xerogels such as pore volume, BET surface area, and mean pore radius were derived from the isotherm analysis. Variations in nanostructures of xerogels with water/TMOS ratio and biomolecule load are shown in Tables 1 and 2, respectively.

TABLE 1. The effect of water/TMOS molar ratio on the surface area (SA), pore volume (PV) and pore radius (PR) of silica xerogels

Ratio	SA, m^2/g	PV, cc/g	PR, nm
4	519	0.21	0.9
6	761	0.31	0.92
10	822	0.35	0.93

TABLE 2. The effect of incorporation of vancomycin (V) and trypsin inhibitor (TI) on the physical properties of silica xerogels

Load/sample	SA, m2/g	PV, cc/g	PR, nm
1 mg V	868	0.39	0.98
10 mg V	936	0.48	1.04
2 mg TI	845	n/d	0.96

• n/d: not determined
• vancomycin load/ sample of 1 and 10 mg correspond to 1.1 and 11 mg/g concentration;
• trypsin inhibitor load of 2 mg corresponds to 6 mg/g concentration.

As shown in Table 1, variation of the water/TMOS ratio does not affect the pore size: PR of xerogels with the ratio of 4, 6 or 10 was about 0.9 nm. However, this variation largely affects the surface area (SA) and pore volume (PV): there is a major reduction in the SA and in the PV values with a decrease of the ratio from 10 to 4.

Concerning the incorporation of bioactive molecules, the presence of vancomycin or trypsin inhibitor did not produce a significant effect on the physical properties of xerogels (Table 2). The effect of other parametrical variations (such as the use of methanol as a solvent, changes in pH from 1 to 5, and variations in the aging and drying schedule) was not significant either.

Therefore, it was found that the nanoporosity of xerogels could be varied to a large degree by altering the synthesis parameters such as the water/alkoxide molar ratio.

2.2. CONTROLLED RELEASE OF DRUGS AND PROTEINS FROM SILICA XEROGELS

Xerogels with various concentrations of drugs (naltrexone or vancomycin) or proteins (trypsin inhibitor), were used in the vitro release study. Xerogels with naltrexone were TEOS-derived, while those with vancomycin or trypsin inhibitor were TMOS-derived. Naltrexone was incorporated in xerogel with a water/alkoxide ratio of 6 (R6 xerogel), whereas vancomycin and the trypsin inhibitor were incorporated in R10 xerogels. Xerogel samples used for the study were shaped either as discs or granules. The elution studies were conducted in tris buffered simulated physiological solution (SPS, pH 7.3 at 37^0C) and solutions were exchanged daily. The ratio of each specimen weight to solution volume was large enough to reach solution saturation with Si-species shortly after immersion and thereby prevent degradation of the samples by dissolution [31]. The immersed samples were maintained in a water-jacketed incubator at 37^0C.

The released concentrations of naltrexone and vancomycin were measured using a UV-visible spectrophotometer (Ultrospec Plus, Pharmacia LKB) at 283 and 280 nm, respectively. A gold colloidal assay (Integrated Separation Systems) was used for the determination of concentrations of trypsin inhibitor. Wet chemical analysis (atomic absorption spectrophotometry, Perkin-Elmer 5100 PC, Norwalk, CT) was used to confirm that solution saturation with Si-concentrations was reached within minutes after immersion of xerogel samples .

<u>Controlled release of drugs (naltrexone and vancomycin)</u>

As shown in Figure 1, a sustained release of naltrexone was observed over 12 days. In this figure, the mean cumulative release of naltrexone from R6 xerogels is plotted as a function of elution time and load (3, 6 and 9 mg/sample), The data also demonstrate that the release rate and the amount released were load-dependent. However, the total recovered percentage did not depend on the load and was about 80% for the various load groups.

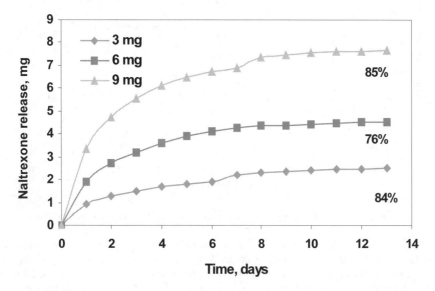

Figure 1. Cumulative release of naltrexone from TEOS-derived xerogel as a function of elution time and load.

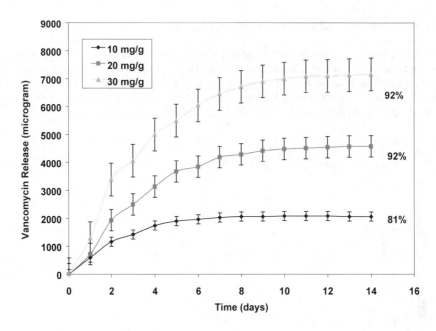

Figure 2. Cumulative release of vancomycin from TMOS-derived xerogel as a function of elution time and load.

The data in Figure 2 demonstrate that the release of vancomycin from R6 xerogels, loaded with 10, 20 or 30 mg/g, was also a controlled, load-dependent, long-term release. The release rates (determined at the linear portions of the release profiles) were 12.4, 21.6, and 34.2 μg/h for the 10-, 20-, and 30-mg/g groups, respectively. These values of hourly release greatly exceeded the minimum inhibitory concentration (MIC) of vancomycin for Gram-positive bacteria (4 μg/ml [32]). By 14 days of immersion, the total release from the 10-mg/g xerogel was about 80% of the original vancomycin load, whereas about 90% was released from 20- and 30-mg/g xerogels.

The possible mechanism of release can be ascertained by presenting the release data as a function of the square root of time [21,23]. Cumulative release of vancomycin from R10 xerogel is presented this way in Figure 3.

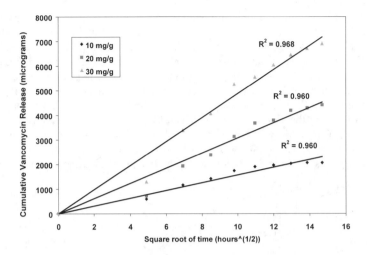

Figure 3. Cumulative release of vancomycin from TMOS-derived R10 xerogel as a function of the square root of time and load.

The linearity of the release plots versus the square root of time reveals first-order release kinetics. As determined by regression analysis, there was no delay and no "burst" in the vancomycin release. Similar presentation of the cumulative release of naltrexone also revealed first-order release without any delay or "burst".

It is well-known that the release which is proportional to the square root of time results from a diffusion driven process. The Highuchi equation relates the total amount of drug released via diffusion from a porous solid matrix into a medium acting as a perfect sink [33]:

$$Q = [(D\varepsilon/\tau) \cdot (2A - \varepsilon C_s) \cdot (C_s t)]^{1/2} \quad (1)$$

where Q is the amount of drug released per unit exposed area after time t, D is the diffusivity of the drug in the permeating fluid, τ is the tortuosity factor of the capillary channel network, A is the total amount of drug present in the matrix per unit volume, C_s is the solubility of the drug in the permeating fluid, and ε is the porosity of the matrix.

The first-order release kinetics of both naltrexone and vancomycin (from R6 and R10 xerogels, respectively) suggests a diffusive mechanism. The model also suggests that the release kinetics depend on the total amount of the drug in the porous matrix. The load-dependent release kinetics of naltrexone and vancomycin agree with the model.

Controlled release of model protein (trypsin inhibitor)

The mean cumulative release of trypsin inhibitor (TI) from R10 xerogels loaded with 2, 5, and 10 mg/sample as a function of elution time is given in Figure 4. The data show a sustained release of TI over a long-term elution period, up to 63 days (9 weeks). Similarly to the release of smaller drug molecules, the release rate of TI and the amount released were load-dependent. However, in contrast to smaller molecules, the TI release was significantly slower. The total release of vancomycin was about 90% by three weeks, whereas only 7, 20 and 15% of TI was released by this time from the 2, 5 and 10 mg load groups, respectively. Also, the pattern of TI release was different from that of the smaller molecules. The first-order release of both naltrexone and vancomycin occurred without any delay, whereas a significant delay in the TI release was revealed by plotting the data against the square root of time [23]. This delay was load-dependent: 24 or 100 hours for the 10 or 2 mg load groups, respectively. This initial delay was followed by first-order release. By 9 weeks, the total recovery was 21, 43 and 32% for the 2, 5 and 10 mg load groups, respectively.

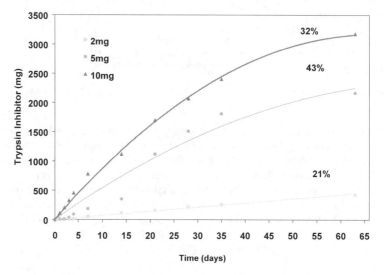

Figure 4. Cumulative release of trypsin inhibitor from TMOS-derived xerogel as a function of elution time and load.

Nanostructural control of release properties

The effect of xerogel nanostructure on the release properties was studied by using xerogels with various water/TMOS molar ratios. As shown in Table 1, variations in the ratio affect the surface area and the pore volume of xerogels without changing their pore size.
Xerogels with the ratio of 4, 6, and 10 (R4, R6, and R10 xerogels) and vancomycin load of 20 mg/g were used. The data in Figure 5 demonstrate a major effect of the ratio on the kinetics of vancomycin release.

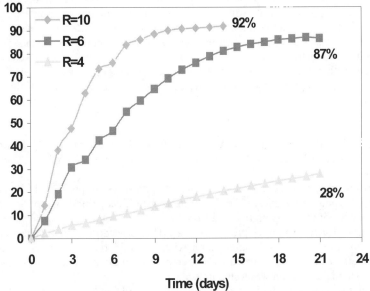

Figure 5. Percent of vancomycin released from TMOS-derived 20-mg/g xerogel as a function of elution time and water/TMOS ratio (R).

The release rates decreased with a decrease in the ratio: the rates of 21.6, 12.2, and 5 μg/h were observed for the R10, R6, and R4 xerogels, respectively (the rates for R10 and R6 xerogels were determined at the linear portions of the plots). About 90% of the vancomycin load was released from R10 and R6 xerogels by 14 and 21 days, respectively. In comparison, only 28% of the load was released from R4 xerogels by 21 days of immersion. Variations in the ratio produced not only a change in the kinetics of release, but also a complete change in the pattern of release. The release from R10 and R6 xerogels was characterized by a first-order release followed by slower, steady release. In comparison, the release from R4 xerogel was a steady, near-zero-order release. This change from a first-order release to a near-zero-order release suggests a change from a diffusion-controlled process to a surface-controlled process.

The observed effect of the water/alkoxide ratio on the release properties is associated with changes in the surface area (SA) and pore volume (PV) (Table 1). In fact, R10 and R6 xerogels with SA above 750 cm^2 and PV greater than 0.32 cc/g, showed a first-order release and a high drug recovery. A major reduction in the SA and PV of R4 xerogels resulted in a major change in the release kinetics of Vancomycin: a slower, zero-order release was observed. These observations suggest a correlation between the porosity of xerogels and the release kinetics. Such a correlation is in agreement with the Highuchi model for the release of a drug from a porous matrix, where the porosity along with other parameters, such as the solubility, diffusivity, and the amount of the drug in the matrix, affect the release kinetics.

The findings of this study suggest that the long-term release properties of silica xerogels can be tailored via nanostructural control. Desirable structural properties can be selected in view of therapeutic requirements. Control of the structural properties via sol-gel processing is simple and does not produce any adverse effect on the biological activity of the therapeutic agents.

2.3. BIOLOGICAL ACTIVITY OF AGENTS RELEASED FROM SILICA XEROGEL: BACTERICIDAL EFFICACY ASSAY OF RELEASED VANCOMYCIN

Vancomycin is the most effective antibiotic against Gram-positive bacteria such as *Staphylococci* and *Streptococci* [32]. The bactericidal efficacy (inhibition of *Staphylococcus aureus* growth) of vancomycin released from the xerogel samples, was assayed according to the standard disc-susceptibility procedure [34]. This procedure includes the steps of agar plate inoculation, paper disk impregnation with an antibiotic solution, and testing of the inhibitory effect of the impregnated disc when placed on the inoculated plate. Each elution sample or standard solution was tested in triplicate.

All the elution solutions (collected at 0.5, 1, 3, 7, 14, and 21 days) showed a strong inhibitory effect (Figures 6 and 7). The zones of inhibition are plotted either as a function of release time (Fig. 6) or as a function of vancomycin concentration in solution (Fig. 7). Figure 7 also shows the efficacy of released vancomycin in comparison to that of standard vancomycin solutions. The zones of inhibition significantly exceeded the zone of the MIC of vancomycin for *S. aureus* (4 µg/ml [32]). Focusing on the semi-logarithmic plots in Figure 7, there is a linear relationship between the dimension of the zone of inhibition and the vancomycin concentration for both solutions tested (the elution solutions and the solutions with calibrated additions of vancomycin). The correlation coefficients, r^2, are equal to 0.995 and 0.999, respectively. This linear relationship is expected in a disc susceptibility test [34]. The overlapping nature of the two sets of data of Figure 7 suggests that the bactericidal efficacy of released vancomycin corresponds to that of the vancomycin solution.

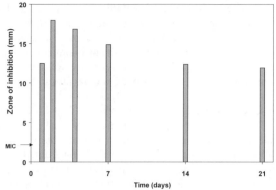

Figure 6. Zone of S. *aureus* inhibition as a function of vancomycin elution time.

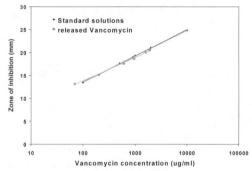

Figure 7. Zone of inhibition of vancomycin released from silica xerogel versus that of standard solutions.

These results provide strong evidence that vancomycin released from silica xerogels fully retains its bactericidal properties. For all molecules studied, we have found that sol-gel processing does not adversely affect the biological activity of molecules incorporated into the xerogels.

2.4. IN VIVO COMPATIBILITY AND RESORPTION BEHAVIOR OF SILICA XEROGELS

The tissue response to silica xerogels was determined in the sub-acute implantation phase (up to 4 weeks of implantation) [26]. We correlated the findings to the composition and the resorption rate of the various xerogels. Silica xerogels with and without vancomycin were used.

All xerogels were synthesized by acid-catalyzed hydrolysis of TMOS. Xerogel implants were shaped either as discs, 8 mm in diameter and 2 mm thick, or as granules in the size range of 710-1000 μm. The materials were sterilized by γ-radiation.

70

Xerogel discs were implanted subcutaneously into the back of New Zealand White rabbits. Additionally, granules were implanted into cylindrical defects (5 mm in diameter and 2 mm in depth) created in the iliac crest. Experimental groups also included controls (sham surgery without implant material). The samples with surrounding tissue were retrieved after either 2 or 4 weeks of implantation. The tissue response was analyzed on thin, stained sections using light microscopy. A total of 34 bone and 38 subcutaneous samples were used for the analysis.

Inflammatory response was scored using a scale from 0 to 4 (0: absent, 1: minimal, 2: mild, 3: moderate, 4: severe) [7]. Morphometric measurements of bone growth and granule size as a function of implantation time were made using a semiautomatic image analysis system consisting of a high resolution color video camera and Image-Pro Plus analysis software. Statistical analysis was performed using a two-way analysis of variance (ANOVA).

Histological analysis of tissue samples indicated that all subcutaneously implanted discs were encapsulated by a pseudo-synovial membrane of densely packed collagen fibers after 2 weeks of implantation. Xerogel discs showed a minimal inflammatory response, after both 2 and 4 weeks of implantation (the score varied from 0.5 to 1.5).

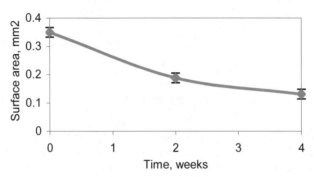

Figure 8. Mean surface area of xerogel granules as a function of implantation time.Error bars represent standard deviation (n=9).

All granules implanted into bone defects showed a gradual decrease in size with implantation time as the morphometric data illustrates in Figure 8. The granule resorption produced a minimal inflammatory response (the score varied from 0.5 to 1.5). The score was lower for granules with vancomycin than for those without it. This resorption was accompanied by extensive trabecular bone ingrowth. As illustrated by light microscopy microphotographs in Figures 9a and 9b, the bone ingrowth for both the control and the xerogel groups was comparable. The trabecular growth was observed in close vicinity to the granules. The trabeculae were covered with a layer of osteoid tissue and a row of active osteoblasts. The healing of the bone defects via trabecular bone growth, evaluated quantitatively as a ratio of new bone to defect area (Figure 10), was statistically the same for the control and material-filled defects at both 2 and 4 weeks of implantation.

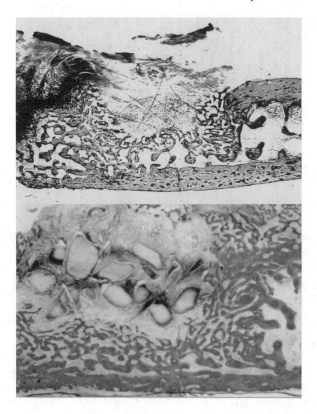

Figure 9. Micrographs of the bone defects after two weeks of implantation: (a) control and (b) defect with implanted xerogel granules. Extensive trabecular bone ingrowth was observed for both the control and the implant groups.

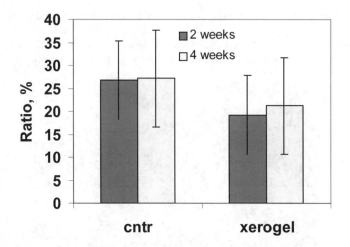

Figure 10. Percent of new bone in the defect as a function of implantation time and experimental groups (control and xerogel). Error bars represent standard deviation (n=9).

This study demonstrates a favorable tissue response to silica xerogels in both subcutaneous and bone sites. Xerogel materials in a granular form showed a gradual resorption, which was accompanied by extensive bone growth. Based on these observations, controlled release silica xerogels are resorbable and biocompatible materials.

3. Conclusion

Based on the in vitro and the in vivo analyses, silica xerogels can be characterized as resorbable and biocompatible materials for the controlled release of drugs and larger biologically active molecules.

References

1. Lewis, D.H. (1990) Controlled release of bioactive agents from lactide/glycolide polymers, in M. Chasin, R. Langer (eds.), *Biodegradable polymers as drug delivery systems*, Marcel Dekker, New York, pp. 1-41.
2. Chasin, M. (1995) Biodegradable polymers for controlled drug delivery, in J.O. Hollinger (eds.) *Biomedical applications of synthetic biodegradable polymers*, CRC Press, Boca Raton, FL, pp. 1-17.
3. Heller, J., Sparer, R.V., Zentner, G.M, in M. Chasin, R. Langer (eds.), *Biodegradable polymers as drug delivery systems*, Marcel Dekker, New York.
4. Leong, K.W., Brott, B.C., and Langer, R. (1985) Bioerodible polyanhydrides as drug-carrier matrices. I. Characterization, degradation, and release characteristics, *J. Biomed. Mater. Res.* **19**, 941-955.

5. Laurencin, C.T., Koh, H.J., Neenan, T.X., Allcock, H.R., and Langer, R. (1987) Controlled release using a new bioerodible polyphosphazene matrix system, *J. Biomed. Mater. Res.* **21**, 1231-1246.

6. Gombotz, W.R., Pankey, S.C., Bouchard, L.S., Phan, D.H., and Puolakkainen, P.A. (1994) Stimulation of bone healing by transforming growth factor-beta$_1$ released from polymeric or ceramic implants, *J. Appl. Biomat.* **5**, 141-150.

7. Royals, M.A., Fujita, S.M., Yewey, G.L., Rodriguez, J., Schultheiss, P.C., Dunn, R.L. (1999) Biocompatibility of a biodegradable in situ forming implant system in rhesus monkeys, *J. Biomed. Mater. Res.* **45**, 231-239.

8. Laurencin, C., Domb, A., Morris, C., Brown, V., Chasin, M., McConnel, R., Lange, N., and Langer, R. (1990) Poly(anhydride) administration in high doses in vivo: Studies of biocompatibility and toxicology, *J. Biomed. Mater. Res.* **24**, 1463-1481.

9. Ibim, S.M., Uhrich, K.E., Bronson, R., El-Amin, S.F., Langer, R.S., Laurencin, C.T. (1998) Poly(anhydride-co-imides): in vivo biocompatibility in a rat model, *Biomaterials* **19**, 941-951.

10. Cornell, C.N., Tyndall, D., Waller, S., Lane, J.M., and Brause, B.D. (1993) Treatment of experimental osteomyelitis with antibiotic-impregnated bone graft substitute, *J. Orthop. Res.* **11**, 619- 626.

11. Mackey, D., Varbet, A., Debeaumont, D. (1982) Antibiotic loaded plaster of Paris pellets: an in vitro study of a possible method of local antibiotic therapy in bone infection, *Clin. Orthop.* **167**, 263-268.

12. Shinto, Y., Uchida, A., Korkusuz, F., Araki, N., and Ono, K. (1992) Calcium hydroxyapatite ceramic used as a delivery system for antibiotics, *J. Bone Surg. (Br)* **74**, 600-604.

13. Yamamura, K., Iwata, H., and Yotsuyanagi, T. (1992) Synthesis of antibiotic-loaded hydroxyapatite beads and in vitro drug release testing, *J. Biomed. Mater. Res.* **26**, 1053-1064.

14. Otsuka, M., Matsuda, Y., Kokubo, T., Yoshihara, S., Nakamura, T., and Yamamura, T. (1995) Drug release from a novel self-setting bioactive glass bone cement containing cephalexin and its physicochemical properties, *J. Biomed. Mater. Res.* **29**, 33-38.

15. Kimakhe, S., Bohic, S., Larrose, C., Reynaud, A., Pilet, P., Giumelli, B., Heymann, D., Daculsi, G. (1999) Biological activities of sustained polymixin B release from calcium phosphate biomaterial prepared by dynamic compaction: An in vitro study, *J. Biomed. Mater. Res.* **47**, 18-27.

16. Benoit, M.A., Mousset, B., Bouillet, R., Gillard, J. (1997) Antibiotic-loaded plaster of Paris implants coated with poly lactide-co-glycolide as a controlled release delivery system for the treatment of bone infections, *Internat. Orthop.* **21(6)**, 403-408.

17. Brinker, C.J., Scherer, G.W. (1990) *Sol-gel Science. The physics and chemistry of sol-gel processing*, Academic Press, San Diego, USA.

18. Braun, S., Rappoport, S., Zusman, R., Avnir, D. and Ottolenghi, M. (1990) Biologically active sol-gel glasses: trapping of enzymes, *Materials Letters* **10(1,2)**, 1-5.

19. Pope, E.J.A. (1995) Living ceramics, *J. Sol-Gel Sci. Tech.* **4**, 225-229.

20. Pope, E.J.A., Braun, K.P., Peterson, C.M. (1997) Bioartificial organs 1: Silica gel encapsulated pancreatic islets for the treatment of diabetes, *J. Sol-Gel Sci. Tech.* **8**, 635-639.

21. Bottcher, H., Slowik, P., Suss, W. (1998) Sol-gel carrier systems for controlled drug delivery, *J. Sol-gel Sci. Tech.* **13**, 277-81.

22. Nicoll, S.B., Radin, S., Santos, E.M., Tuan, R.S., Ducheyne, P. (1997) In vitro release kinetics of biologically active transforming growth factor-β1 from novel porous glass carrier, *Biomaterials* **18**, 853-859.

23. Santos, E,M,, Radin, S., Ducheyne, P. (1999) Sol-gel derived carrier for the controlled release of proteins, *Biomaterials* **20**, 1695-1700.

24. Radin, S., Ducheyne, P., Kamplain, T., Tan, B.H. (2001) Silica sol-gel for the controlled release of antibiotics. I. Synthesis, characterization, and in vitro release, *J. Biomed. Mater. Res.* **57**, 313-320.

25. Aughenbaugh, W., Radin, S., Ducheyne, P. (2001) Silica sol-gel for the controlled release of antibiotics. II. The effect of synthesis parameters on the in vitro release kinetics of vancomycin, *J. Biomed. Mater. Res.* **57**, 321-326.

26. Radin, S., El-Bassyouni, G., Vresilovic, E.J., Schepers, E., Ducheyne, P. In vivo tissue response to resorbable silica xerogels as controlled release materials, *Biomaterials*, submitted

27. Ahola, M., Kortesuo, P., Kangasniemi, I., Kiesvaara, J., Yli-urpo, A. (2000) Silica xerogel carrier material for controlled release of toremifene citrate, *Int. J. Pharm.* **195**, 219-27.

28. Kortesuo, P., Ahola, M., Kangas, M., Kangasniemi, I., Yli-Urpo, A., Kiesvaara, J. (2000) In vitro evaluation of sol-gel processed spray dried silica gel microspheres as carrier in controlled drug delivery, *Int. J. Pharm.* **200**, 223-9.

29. Brunauer, S., Emmett, P. and Teller, E. (1938) *J. Amer. Chem. Soc.* **60**, 309.

30. Brunauer, S., Deming, L.S., Deming, W.S. and Teller, E. (1940) *J. Amer. Chem. Soc.* **62**, 1723.

31. Falaize, S., Radin, S., and Ducheyne, P. (1999) In vitro behavior of silica-based xerogels intended as controlled release carriers, *J. Amer. Chem. Soc.* **82(4)**, 969-976.

32. Geraci, J.E., Hermans, P.E. (1983) Vancomycin, *Mayo Clin. Proc.* **58**, 88-91.

33. Highuchi, T. (1963) Mechanism of sustained-action medication: Theoretical analysis of rate of release of solid drugs dispersed in solid matrices, *J. Pharm. Sci.* **52**, 1145-1149.

34. Acar, J.F. and Golstein, F.W. (1991) Disk susceptibility test, in V.L. Lorian (eds.), *Antibiotics in Medicine. Baltimore*, Williams and Wilkins, pp.17-52.

SURFACE ANALYSIS OF BIOMATERIALS AND BIOMINERALIZATION

BUDDY D. RATNER
Center for Bioengineering and Department of Chemical Engineering
University of Washington, Box 351750
Seattle, Washington 98195 USA

1. Introduction

The atoms and molecules that reside at the surface of all condensed matter have a special organization and reactivity. Surface atoms and molecules require special methods to characterize them, novel methods to tailor them and they drive many of the biological reactions that occur in response to the biomaterial (protein adsorption, cell adhesion, cell growth, blood compatibility, biomineralization etc.). Surface characterization for the study of biomaterials has been appreciated as an important topic since the 1960's and almost every biomaterials meeting will have sessions addressing surfaces and interfaces. In this chapter we focus on the special methods to characterize surfaces and some implications of surfaces for bioreaction to biomaterials.

During the development of a medical device or material we are concerned with function, durability, and biocompatibility. In order to function, the implant must have appropriate physical properties such as mechanical strength, permeability or elasticity, just to name a few of these properties. Well-developed methods exist to measure these bulk properties – often these are the classic methodologies of engineers and materials scientists. Durability, particularly in a biological environment, is less well understood. Still, the tests we need to assess durability have been developed over the past 20 years. Biocompatibility represents a frontier of knowledge in this field, and its study is often assigned to the biochemist, biologist, and physician. However, an important question in biocompatibility is how the device or material "communicates" its structural makeup to direct or influence the response of proteins, cells, and the organism to it. For devices and materials that do not leach undesirable substances (i.e., that have passed routine toxicology), this "communication" occurs through the surface structure -- the body "reads" the surface structure and responds. Thus, we must understand biomaterial *surface* structure.

R.L. Reis and S. Weiner (eds.),
Learning from Nature How to Design New Implantable Biomaterials, 75-85.
© 2004 *Kluwer Academic Publishers. Printed in the Netherlands.*

Surface analysis methods were developed largely in the solid-state physics and microelectronics communities. They are relatively new to biology. A compendium of these methods is offered in Table 1.

TABLE 1. Methods to Analyze Surfaces

Method	Acronym	Depth	Probe	Measurement
Contact Angles	-	-3-20Å	liquid	contact angle
Electron Spectroscopy for Chemical Analysis	ESCA (also, XPS)	15-100Å	X-rays	electrons
Auger Electron Spectroscopy*	AES (also, SAM)	50-100Å	Electrons	Electrons
Surface Extended X-ray Absorption Fine Structure	SEXAFS	50-100Å	X-rays (tunable)	electrons
Secondary Ion Mass Spectrometry	SIMS	10Å-1μm**	ions	ions (mass)
Ion Scattering Spectroscopy	ISS	5Å	ions	ions (energy)
Attenuated Total Reflection Infrared	ATR-IR	1-5μm	infrared	absorbance (vibrations)
High Resolution Electron Energy Loss Spectroscopy	HREELS	5Å	electrons	energy loss (vibrations)
Infrared Reflection Absorption Spectroscopy	IRAS	5Å-100Å	infrared	absorbance
Scanning Tunneling Microscopy	STM	5Å	tip (metal)	tunneling electrons
Atomic Force Microscopy	AFM	5Å	tip (SiN)	interactive force
Scanning Electron Microscopy	SEM	5Å	electrons	electrons

Figure 1 categorizes these techniques into six classes of methods and addresses their sampling depth into materials. The advantages of having many methods available are that each offers different, but complementary, information about a surface, and probes that surface to different depths. This review article will offer insights into just a few of the most generally useful surfaces methods. Additional reference to application for biomineralization will be made.

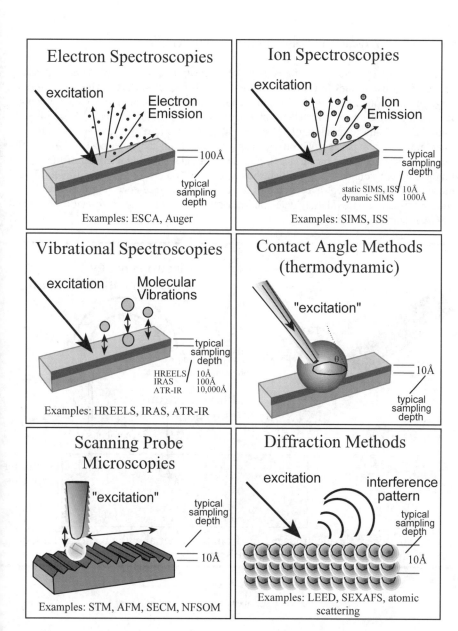

Figure 1. The methods useful for the analysis of solid surfaces can be placed in six general categories.

2. Electron Spectroscopy for Chemical Analysis (ESCA)

The ESCA method (also referred to as x-ray photoelectron spectroscopy, XPS) is possibly the single most useful tool to characterize biomaterials and biosurfaces. This is because of the high information content of ESCA, good commercial instrumentation, a solid theoretical understanding and well-established practice. The principle behind ESCA is the measurement of the energies of electrons emitted from a surface in response to bombardment by soft x-rays. These electrons emerge from the outermost ~100Å of the surface (a useful depth for biosurfaces), and their energies are characteristic of the atomic and molecular environments from which they originated (Figure 2). ESCA will identify and quantify all elements except hydrogen and helium in the surface zone. Furthermore, it will provide information on oxidation states and molecular environments. Information provided by ESCA is summarized in Table 2.

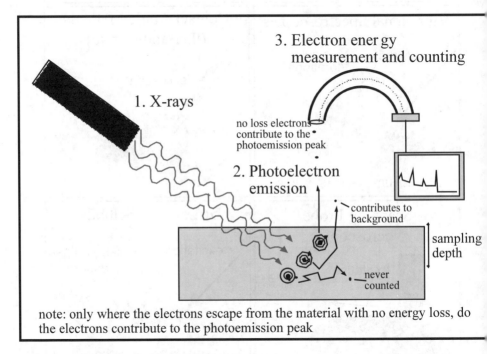

Figure 2. The ESCA method of surface analysis.

TABLE 2. Information derived from an ESCA experiment

In the outermost 100Å of a surface, the following information can be derived from an ESCA experiment:

- Identify all elements (except H and He) present at concentrations >0.1 atomic %
- Semiquantitative determination of the elemental surface composition ($\pm$ 10%)
- Molecular environments (oxidation state, bonding atoms, etc.)
- Aromatic or unsaturated structures from shake-up (π^* - π) transitions
- Identification of specific organic groups with derivatization reactions
- Nondestructive depth profiles (top 100Å) and surface heterogeneity assessment
- Destructive elemental depth profiles >1000Å into the sample with Ar etching
- Spatial resolution from 1-150 μm, depending upon the instrument
- Valence band information and identification of bonding orbitals
- Analysis of hydrated (frozen) surfaces

A number of enhancements have been made to the basic ESCA method that further expand its utility. These include angular dependent studies permitting non-destructive depth profiles of the outermost 100Å; frozen-hydrated experiments for studying the ability of water and solvents to restructure the surface; and chemical post-derivatization studies that allow a more precise assignment and quantitation of specific organic functional groups.

3. Secondary Ion Mass Spectrometry (SIMS)

SIMS permits a mass spectrum of the surface zone to be taken. A surface is bombarded by energetic ions (3-15 KeV). These ions (primary ions) transfer their momentum into the surface zone imparting sufficient energy to some surface atoms, molecules or fragments of molecules that they can be emitted as ions (secondary ions) from the surface (Figure 3). Figure 3 also suggests two possibilities for mass analysis, a quadrupole analyzer and a time-of-flight (ToF) analyzer. After emission, the molecular weight (more accurately, mass over charge, m/z) of the secondary ions is measured. SIMS in the static mode is essential for organic materials to avoid extensive degradation and rearrangement associated with the energetic primary ions. In static SIMS, the ion flux is kept sufficiently low so that over the analysis time, no more than 10% of the surface atoms are eroded away. At the site where the primary ion impacts the surface, massive fragmentation occurs leading to the emission of small ions such as C- (m/z 12), CH- (m/z 13), O- (m/z 16), OH- (m/z 17), Na+ (m/z 23), etc., frequently called "atomics."

Surrounding the impact crater, from relatively undamaged surface, larger, complex, fragment ions are produced that directly code information on surface structure. These are referred to as "moleculars." The static SIMS criterion (no more than 10% of the surface atoms eroded) ensures that most of the impact craters are far apart and maximizes surface emission of the interesting molecular ions. The sampling depth is roughly 10-15Å in the static SIMS mode.

Static SIMS produces secondary ion fragments that typically have a close resemblance to the surface structure. The rules of spectral interpretation are similar to those used for conventional mass spectrometry. In particular, SIMS spectra can be used as "fingerprints" for comparison with spectra in handbooks. Alternately, specific gas phase ions can be identified. Finally, surface structures can be surmised based upon the ions identified. Tandem SIMS experiments can further assist in identifying the nature of gas phase ions.

SIMS analysis can be enhanced with isotope substitutions (particularly effective for biomineralization and plasma depsoition studies), chemical derivatization, post-ionization (to convert neutral emitted particles to measurable ions) and multivariate statistics for pattern recognition and quantitative analysis. Also, rough silver surfaces can be used as deposition substrates for extremely thin organic films leading to intense ion production where each molecular fragment is associated with one silver ion.

Crosslink density and other morphological characteristics of the surface zone might can be probed by SIMS. The ratio of atomic fragments to larger molecular fragments appears sensitive to the chain length between crosslinks, molecular mobility and sample crystallinity. Thus, coded information is contained in the SIMS spectrum in addition to the rich chemical information that SIMS provides.

ToF-SIMS has greatly advanced the instrumentation state-of-the-art over the quadrupole mass analyzers that dominated SIMS just a few years ago. ToF-SIMS allows very high mass resolution (for example, a fragment at nominal mass m/z 43 can be accurately identified as $C_2H_3O^+$ based upon its precise mass, 43.019). Also, extremely high analytical sensitivity and identification of high mass fragments (sometimes greater than $10,000 m/z$) is possible.

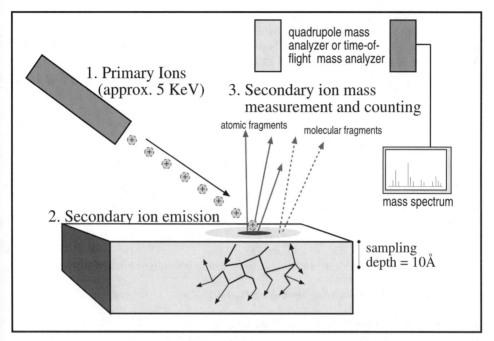

Figure 3. Key aspects of the SIMS method of surface analysis.

4. Atomic Force Microscopy (AFM)

AFM, a tool for observing the topography and mechanics of surfaces, is in the scanning probe microscopy family of methods. AFM is particularly useful for biomaterials because it can be used for hard and soft materials, insulators and conductors. A silicon nitride tip terminating in just a few atoms, attached to a silicon cantilever of extremely low modulus, is passed over a surface using piezoelectric controllers. The deflection of the cantilever is amplified by a laser optical signal to provide a magnified profile of the surface topography. Topographic features on the surface can be characterized over microns, or with molecular scale resolution. Since the cantilever obeys Hooke's Law, the forces of interaction at the nanoscale can be measured -- the mechanical moduli of fine features on the surface can be deduced based upon cantilever deflection. Also, by chemically derivatizing the tip, forces of interaction between surface molecular-scale features and the tip can be measured. With specific receptor groups immobilized on the tip, a "recognition" microcopy can be practiced. The frictional interaction between tip and surface can be measured. Finally, by studying the phase relationships of an oscillating tip in proximity with a surface, insights about the mechanical properties of the surface can be obtained.

Related methods to the AFM are also of interest for surface characterization. The scanning electrochemical microscope (SECM) provides information on ionic conductivity of surfaces at the sub-micron level of resolution. Near field scanning optical microscopy (NFSOM) techniques permit visible light images of surface features at sub-micron resolution.

5. Other Surface Methods

There are many other surface analytical tools that can assist in the elucidation of the structure of biosurfaces. These include contact angles to measure surface energetics and wettability; attenuated total reflectance infrared (ATR-IR -- penetration depths of 1-5μm); infrared reflection absorption spectroscopy (IRAS -- can be used to probe molecular orientation in the outermost ~100Å); ellipsometry for film thickness; near edge x-ray absorption fine structure (NEXAFS) for molecular orientation information; surface plasmon resonance (SPR) to studying adsorption and interaction events in aqueous and real time; and electron energy loss spectroscopy (EELS) for molecular vibrational information at the outermost surface.

Articles in the bibliography can provide details on methods mentioned in this article.

6. Surface Analysis for Studying Biomineralization

The analysis of biomineralization has been dominated by x-ray diffraction for assessing crystalline structure and by microscopy to visualize the mineralization. Vibrational spectroscopies have also been applied to look at aspects of biomineral chemistry and structure. Surface analysis can provide unique insights into biomineral formation, structure and biointeractions. ESCA offers chemical shifts in binding energies that are highly sensitive to bonding environments. Also, ESCA offers quantitative information, so stoichiometry is readily assessed with excellent accuracy. Many of the crystalline phases of calcium phosphates and calcium carbonates can be distinguished by ESCA. Similarly SIMS shows peaks that are often unique, or unique peak patterns that can be correlated with different mineral crystalline phases. This information is, of course, just at the surface zone (outermost 100Å), which is not probed by X-ray diffraction, but is indeed observed by interacting proteins and cells. Surface analysis methods also look at the proteins and biomolecules interacting with biomineral surfaces and can identify possible contaminants. Finally, surface methods can be used in the study of mineralizing implants such as hip joints. The biomineralization section of the bibliography gives key references in this area.

Acknowledgement

Many of the ideas in this article, and some of the research described, came about through research funded by the National Center for Research Resources (NCRR) of the NIH, grant RR01296 (NESAC/BIO) and University of Washington Engineered Biomaterials (UWEB), NSF EEC-952916.

Surface Analysis Bibliography: Selected References

GENERAL

Castner, D.G., Ratner, B.D. (2002) Biomedical surface science: foundations to frontiers, *Surf. Sci.* **500**, 28-60.

Johnston, E., Ratner, B.D. (1996) Surface characterization of plasma deposited organic thin films, *J. Elect. Spectrosc. Rel. Phenom.* **81**, 303-317.

Ratner, B.D., Porter, S.C. (1996) Surfaces in biology and biomaterials: description and characterization, in J. Brash and P. Wojciechowski (eds.), *BioProducts at Interfaces*, Marcel Dekker, New York, pp. 57-83.

Ratner, B.D. (1995) Advances in the analysis of surfaces of biomedical interest, *Surf. Interface Anal.* **23**, 521-528.

Garbassi, F., Morra, M., Occhiello, E. (1994) *Polymer surfaces: from physics to technology.* 1st ed. John Wiley and Sons, Chicheston,UK.

Ratner,BD; Castner,DG (1994): Advances in XPS instrumentation and methodology: Instrument evaluation and new techniques with special reference to biomedical studies. Coll. Surf. B: Biointerfaces 2, 333-346.

Ratner, B.D. (1993) Characterization of biomaterial surfaces, *Cardiovasc. Pathol.* **2** Suppl.(3), 87S-100S.

Briggs, D. Seah, M.P. (1990) *Practical surface analysis.* 2nd ed. Vol. 1. John Wiley and Sons, Chichester.

ESCA

Ratner, B.D., Castner, D.G. (1996) Electron spectroscopy for chemical analysis, in J.C. Vickerman and N.M. Reed (eds.), *Surface Analysis - Techniques and Applications.* 1st ed. John Wiley and Sons, Ltd., Chichester, UK.

Ratner, B.D., McElroy, B.J. (1986) Electron spectroscopy for chemical analysis: applications in the biomedical sciences, in R.M. Gendreau (eds.), *Spectroscopy in the Biomedical Sciences*, CRC Press, Boca Raton, Fl, pp. 107-140.

DERIVATIZATION

Chilkoti, A. Ratner, B.D. (1993) Chemical derivatization methods for enhancing the analytical capabilities of X-ray photoelectron spectroscopy and static secondary ion mass spectrometry, in L. Sabbatini and P.G. Zambonin (eds.), *Surface Characterization of Advanced Polymers*, VCH Publishers, Weinheim, Germany, pp. 221-256.

SYNCHROTRON METHODS

Castner, D.G., Lewis, K.B., Fischer, D.A., Ratner, B.D., Gland, J.L. (1993) Determination of surface structure and orientation of polymerized tetrafluoroethylene films by near-edge X-ray absorption fine structure, X-ray photoelectron spectroscopy, and static secondary ion mass spectrometry. *Langmuir* **9**, 537-542.

Preses, J.M., Grover, J.R., Kvick, A., White, M.G. (1990) Chemistry with synchrotron radiation, *Am. Sci.* **78**, 424-437.

SIMS

Belu, A.M., Graham, D., Castner, D.G. (2003) Time-of-flight secondary ion mass spectrometry: techniques and applications for the characterization of biomaterial surfaces, *Biomaterials* **24**, 3635-3653.

Castner, D.G., Ratner, B.D. (1988) Static secondary ion mass spectroscopy: a new technique for the characterization of biomedical polymer surfaces, in B.D. Ratner (eds.), *Surface Characterization of Biomaterials*, Elsevier Press, Amsterdam, pp. 65-81.

Mantus, D.S., Ratner, B.D., Carlson, B.A., Moulder, J.F. (1993) Static secondary ion mass spectrometry of adsorbed proteins, *Anal. Chem.* **65**, 1431-1438.

Benninghoven, A. (1994) Chemical analysis of inorganic and organic surfaces and thin films by static time-of-flight secondary ion mass spectrometry (TOF-SIMS), *Angew. Chem. Int. Ed. Engl.* **33**, 1023-1043.

Davies, M.C., Lynn, R.A.P. (1990) Static secondary ion mass spectrometry of polymeric biomaterials, *CRC Crit. Rev. Biocompat.* **5(4)**, 297-341.

Chilkoti, A., Ratner, B.D., Briggs, D. (1991) A static secondary ion mass spectrometric investigation of the surface structure of organic plasma- deposited films prepared from stable isotope-labeled precursors part I. Carbonyl precursors, *Anal. Chem.* **63**, 1612-1620.

Chilkoti, A., Ratner, B.D., Briggs, D. (1993) Static secondary ion mass spectrometric investigation of the surface chemistry of organic plasma- deposited films created from oxygen-containing precursors. 3. Multivariate statistical modeling, *Anal. Chem.* **65**, 1736-1745.

Briggs, D., Hearn, M.J., Ratner, B.D. (1984) Analysis of polymer surfaces by SIMS 4 - A study of some acrylic homo- and co-polymers, *Surf. Interface Anal.* **6(4)**, 184-192.

Chilkoti, A., Ratner, B.D., Briggs, D. (1992) Analysis of polymer surfaces by SIMS: Part 15. Oxygen-functionalized aliphatic homopolymers, *Surf. Interface Anal.* **18**, 604-618.

HYDRATED-FROZEN SURFACE ANALYSIS

Lewis, K.B., Ratner, B.D. (1993) Observation of surface restructuring of polymers using ESCA, *J. Coll. Interf. Sci.* **159**, 77-85.

VIBRATIONAL SPECTROSCOPIES

Yates, J.T.Jr. and Madey, T.E (1987) *Vibrational Spectroscopy of Molecules on Surfaces*, Plenum Press, New York.

Harrick, N.J. (1967) *Internal Reflection Spectroscop,*. Interscience Publishers, New York.

AFM

Jandt, K.D. (2001) Atomic force microscopy of biomaterials surfaces and interfaces, *Surf. Sci.* **491**, 303-332.

Luginbuhl, R., Overney, R.M., and Ratner, B.D. (2001) Nanobiotribology at the Confined Biomaterial Interface. in J. Frommer, R.M. Overney (eds.), *Interfacial Properties on the Submicrometer Scale*, American Chemical Society, Washington, DC, pp. 178-196.

Newman, A. (1996) Beyond the surface: Looking at atomic force microscopes, *Anal. Chem.* (April 1, 1996), 267A-273A.

Boland, T. and Ratner, B.D. (1995) Direct measurement by atomic force microscopy of hydrogen bonding in DNA nucleotide bases, *Proc. Natl. Acad. Sci. USA* **92(12)**, 5297-5301.

Quate, C.F. (1994) The AFM as a tool for surface imaging, *Surf. Sci.* **299/ 300**, 980-995.

Lindsay, S.M., Lyubchenko, Y.L., Tao, N.J., Li, Y.Q., Oden, P.I., DeRose, J.A., and Pan, J. (1993) Scanning tunneling microscopy and atomic force microscopy studies of biomaterials at a liquid-solid interface, *J. Vac. Sci. Technol. A* **11(4)**, 808-815.

BIOMINERALIZATION SURFACE STUDIES

Barbotteau,Y., Irigaray, J.L., and Jallot, E. (2003) Physicochemical characterization of biological glass coatings, *Surface and Interface Analysis* **35 (5)**, 450-458.

Feddes, B., Vredenberg, A.M., Wolke, J.G.C., and Jansen, J.A. (2003), Determination of photoelectron attenuation lengths in calcium phosphate ceramic films using XPS and RBS, *Surface and Interface Analysis* **35 (3)**, 287-293.

Jelvestam, M., Edrud, S., Petronis, S., Gatenholm, P. (2003) Biomimetic materials with tailored surface micro-architecture for prevention of marine biofouling, *Surface and Interface Analysis* **35(2)**, 168-173.

Lu, H.B., Campbell, C.T., Graham, D., and Ratner, B.D. (2000) Surface characterization of hydroxyapatite and related calcium phosphates by XPS and TOF-SIMS, *Anal. Chem.* **72(13)**, 2886-2894.

Ni, M. and Ratner, B.D. (2003) Nacre surface transformation to hydroxyapatite in a phosphate buffer solution. *Biomaterials* **24**, 2423-4331.

Jallot, E., Irigaray, J.L., Weber, G., and Frayssinet, P. (1999) In vivo characterization of the interface between cortical bone and biphasic calcium phosphate by the PIXE method, *Surface and Interface Analysis* **27(7)**, 648-652.

MacDonald, D.E., Betts, F., Stranick, M., Doty, S., and Boskey, A.L. (2001) Physicochemical study of plasma-sprayed hydroxyapatite-coated implants in humans, *Journal of Biomedical Materials Research* **54(4)**, 480-490.

Effah Kaufmann, E.A.B., Ducheyne, P., Radin, S., Bonnell, D.A., Composto, R. (2000) Initial events at the bioactive glass surface in contact with protein-containing solutions, *Journal of Biomedical Materials Research* **52(4)**, 825-830.

Kirk, P.B., Filiaggi, M.J., Sodhi, R.N.S., and Pilliar, R.M. (1999) Evaluating sol-gel ceramic thin films for metal implant applications: III. In vitro aging of sol-gel-derived zirconia films on Ti-6Al-4V, *Journal of Biomedical Materials Research* **48(4)**, 424-433.

Joshi, R.R, Frautschi, J.R., Phillips, R.E.Jr., Levy, R.J. (1994) Phosphonated polyurethanes that resist calcification, *J. Appl. Biomat.* **5**, 65-77.

Lodding, A.R., Fischer, P.M., Odelius, H., Noren, J.G., Sennerby, L., Johansson, C.B., Chabala, J.M., and Levi-Setti, R. (1990) Secondary ion mass spectrometry in the study of biomineralizations and biomaterials, *Anal. Chim. Acta* **241**, 299-314.

Porter, S.C. and Ratner, B.D. (1997) Investigation of cation/polyether complexes with static ToF-SIMS; Transactions of the 23rd Annual Meeting of the SOCIETY FOR BIOMATERIALS, April 30-May 4, 1997, New Orleans, Louisiana, USA.

3. Biomimetics and Biomimetic Coatings

BIOMIMETICS AND BIOCERAMICS

B. BEN-NISSAN
University of Technology, Sydney
Department of Chemistry, Materials and Forensic Science
PO BOX 123, Broadway 2007, NSW
AUSTRALIA
E-mail: b.ben-nissan@uts.edu.au

Abstract

The synthesis of complex inorganic forms, which are based on natural structures that can mimic the natural scaffold upon which the cells are seeded, offers an exciting range of avenues for the construction of a new generation of bone analogs for tissue engineering. The production and use of synthetic calcium phosphate bioceramics based on the coralline structure as bone grafts in orthopedics is considered. Recent advances in the production and use of natural bioceramics for application in hard and soft tissue replacements are discussed. An improved understanding of currently used bioceramics in human implants and in bone replacement materials could contribute significantly to the design of new generation prostheses and post-operative patient management strategies. Issues affecting the use of different materials *in vivo* are outlined. A variety of other natural alternatives including newly developed sol-gel coated coralline apatite is evaluated, and reviewed. Several treatments for improving performance are outlined, and speculation on future advances, including the combination of traditional bioceramic implants with natural biogenic additives is made.

Keywords: bioceramics, biomimetics, sol-gel, bone-grafts, calcium phosphates

1. Introduction

Throughout life, the human body's muscular-skeletal system is constantly being formed and resorbed; with increasing age this leads to a reduction in bone mass and density and a need to replace or repair the degenerated bone and restore its biomechanical and biologic function. Techniques to combat this serious problem have now become a major clinical need. Currently the worldwide biomaterials market is valued at close to US$24,000M. Orthopedic and dental applications represent approximately 55% of the total biomaterials market. Orthopedic products worldwide exceeded $13,000M in 2000, an increase of 12 percent over 1999

R.L. Reis and S. Weiner (eds.),
Learning from Nature How to Design New Implantable Biomaterials, 89-103.
© 2004 *Kluwer Academic Publishers. Printed in the Netherlands.*

revenues. Expansion in these areas is expected to continue due to a number of factors, including the ageing population, an increasing preference by younger to middle-aged candidates for undertaking surgery, improvements in technology and lifestyle, a better understanding of body functionality, improved esthetics and a need for better function [1].

Some of the earliest biomaterial applications occurred as far back as ancient Egypt and Phoenicia, where loose teeth were bound together with gold wires for tying artificial ones to neighbouring teeth. From as early as 19th century, artificial materials and devices have been developed to a point where they can replace various components of the human body. These materials are capable of being in contact with bodily fluids and tissues for prolonged periods of time, whilst eliciting little if any adverse reaction [2]. In the early 1900s bone plates were successfully implemented to stabilize bone fractures and to accelerate their healing. By the 1950s to 1960s, blood vessel replacements were being clinically trialed and artificial hip joints were in development.

Even in the preliminary stages of this field's development, engineers and surgeons identified materials and design problems that resulted in premature loss of implant function through inadequate biocompatibility of the component, mechanical failure, raw material and synthesis problems, wear, degradation or corrosion. The key factors in biomaterial usage are biocompatibility and biofunctionality under various functional loadings. Ceramics are ideal candidates with respect to all the above functions, except for their brittle behavior.

2. Bioceramics

It has been widely accepted that the only substances that conform completely to the human body's environment are those manufactured by the body itself (autogenous) and any other substance that is recognized as foreign, and initiates some type of reaction.

When a synthetic material is placed within the body, tissue reacts towards the implant in a variety of ways, depending on the material type. The mechanism of tissue interaction depends on the tissue's response to the implant surface. In general, a biomaterial may be described in or classified into representing the tissue's responses in three terms. These are bioresorbable, bioactive and bioinert [3-5].

Bioresorbable refers to a material that, upon placement within the human body, starts to dissolve and is slowly replaced by advancing tissue. Common examples of bioresorbable materials are tricalcium phosphate [$Ca_3(PO_4)_2$] and polylactic–polyglycolic acid copolymers. Calcium oxide, calcium carbonate (coral) and gypsum are other common materials that have been utilized during the last three decades.

Bioactive refers to a material, which, upon being placed within the human body, interacts with the surrounding bone and, in some cases, even soft tissue. An ion exchange reaction between the bioactive implant and surrounding body fluids in some cases results in the formation of a biologically active carbonate apatite (CHA) layer on the implant that is chemically and crystallographically equivalent to the mineral phase of bone. Prime examples of these materials are synthetic hydroxyapatite [6,7] [$Ca_{10}(PO_4)_6(OH)_2$], glass-ceramic A-W [8,9] and bioglass® [10].

The term bioinert refers to any material that, once placed within the human body, has minimal interaction with its surrounding tissue. Generally, a fibrous capsule might form around bioinert implants; hence its biofunctionality relies on tissue integration through the implant. Examples of these bioinert materials are stainless steel, titanium, cobalt-chromium molybdenum alloy (Zimmer alloy), alumina, partially stabilized zirconia, new generation zirconia and allumina alloys and ultra-high molecular weight polyethylene.

Interest in ceramics for biomedical applications has increased over the last thirty years. The ceramics that are used in implantation and clinical purposes include aluminium oxide (alumina), partially stabilized zirconia (PSZ) (both yttria [Y-TZP] and magnesia stabilized [Mg-PSZ]), bioglass®, glass-ceramics, calcium phosphates (hydroxyapatite and β-tricalcium phosphate) and crystalline or glassy forms of carbon and its compounds [11].

Ceramics are considered hard, brittle materials with excellent compressive strength, high resistance to wear, and with favourably low frictional properties in articulation, however, with relatively poor tensile properties. The low frictional properties are enhanced by the fact that ceramics are hydrophilic with good wettability and can be highly polished, which provides a superior lubrication on a load-bearing surface with itself or against polymeric materials in a physiologic environment. Bioceramics used singularly or with additional natural, organic or polymeric materials are amongst the most promising of all synthetic biomaterials for hard and soft tissue applications.

3. Bioresorbable and Bioactive Ceramics

The first x-ray diffraction study of bone was initiated [12] by De Jong in 1926, in which apatite (dahllite-carbonated apatite) was identified as the only recognizable mineral phase. He also reported marked broadening of the diffraction lines of bone apatite, which he attributed to small crystal size. It was not until the 1970s that synthetic hydroxyapatite [$Ca_{10}(PO_4)_2(OH)_2$] was accepted as a potential biomaterial that forms a strong chemical bond with bone *in vivo*, while remaining stable, under the harsh conditions encountered in the physiologic environment. Since then, research has suggested the existence of other mineral phases in bone, including amorphous calcium phosphate (ACP), brushite, and octacalcium phosphate (OCP). The presence of substantial amounts of ACP or brushite has not been yet experimentally proven, and nuclear magnetic resonance (NMR) studies support the conclusion that bone is composed essentially of carbonate substituted hydroxyapatite (CHA).

Hydroxyapatite (HA) [13] or, more accurately, carbonate hydroxyapatite similar to the dahllite mineral [14], is the major constituent of the human bone. It is now agreed that the bone apatite (like the mineral of normally calcified tissues) can be better described as carbonate hydroxyapatite (CHA), approximated by the formula:

$$(Ca,Mg,Na)_{10}(PO_{4,}CO_3)_6(OH)_2 \quad [15,16,17]$$

Biogenic apatite is believed to be formed from octacalcium phosphate (OCP) via stoichiometric protons loss and substitution of calcium ions for protons (Equation 1) [18,19].

$$Ca_8H_2(PO_4)_6.5H_2O + 2Ca^{2+} \rightleftharpoons Ca_{10}(PO_4)_6(OH)_2 + 4H^+ + 3H_2O \quad (Eq.\ 1)$$

The process is topotactic [20,21], that is, transformation of OCP to apatite with no change of the morphology. During transformation carbonate ions may be incorporated into the apatite crystallites [22-25]. The incorporation may be of two types. When a carbonate anion occupies lattice positions in the OH⁻ sublattice it is named A-type carbonate apatite. Carbonate can also be incorporated via phosphate displacement. This is called B-type substitution. Bone apatite contains CO_3^{2-} in both types of locations in ratio A to B of 0.7 to 0.9 [23,24]. Although both types A and B carbonate apatite in natural bone are widely reported, a recent laser Raman Spectroscopy work showed that no OH- band was detected in natural bone. It was further reported that this observation is unexpected, and it therefore remains unclear what atoms occupy the OH- site and how charge balance is maintained within the crystal [25].

The crystallites of bone apatites are plate shaped nano sized particles with lengths and widths of 50 x 25 nm and thickness of 2 – 5 nm. Hence, synthetic strategies to produce bone apatite confront two challenges: to incorporate carbonate in possibly both A and B positions and to mimic morphology.

Synthetic hydroxyapatite $Ca_{10}(PO_4)_6(OH)_2$ has been an attractive material for chromatographic separation catalysis, ion exchange, and bone and tooth implants [26]. Since its inception one of the most common and easiest production methods of synthetic HAp has been by a solid state reaction between Ca^{2+} and PO_4^{3-} bearing compounds and under solution conditions in the form of powders which can be sintered to a dense polycrystalline body by firing. Parameters such as the Ca/P ratio, purity, grain size and secondary compounds that could form during its production usually control the bioactivity of HAp [27,28].

Albee and Morrison [29] proposed the use of calcium phosphate ceramics for biomedical applications after observing accelerated bone growth, when β-tri-calcium phosphate was injected into bone defects. Pure β-tri-calcium phosphate (β-TCP) $Ca_3(PO_4)_2$ is more soluble in the physiological environment than other phosphate ceramics (bioresorbable). Consequently, it can be used in situations where accelerated bone growth is desirable. β-tricalcium phosphates have been used successfully, as fillers for bone defects to stimulate the formation of new bone [30]. This work also showed that after a 12-month period, β-tricalcium phosphate was totally absorbed. These materials are intended to be used in filling voids in bone structure that will dissolve over a period of time while the dissolution takes place, and the bone re-growth or advancement takes place at similar rates. In current commercial products (β-TCP) and HAp are mixed in predetermined proportions to induce a controlled dissolution rate.

The dissolution rates of some of these materials under simulated physiological conditions have been investigated with emphasis placed on hydroxyapatite, β-tricalcium phosphate and tetra-calcium phosphate. Under *in vitro* conditions, the solubility of these materials has been shown to decrease in the order of [31].

Tetra-calcium phosphate > β-Tricalcium Phosphate > Hydroxyapatite

It was proposed by various investigators that the initial formation of an amorphous calcium phosphate (ACP) at a high pH could be followed by its transformation to hydroxyapatite (HA) via the formation of a precursor in the form of octacalcium phosphate (OCP). This has been proposed to be one of the templates that form HA. It has also been stated that, as the pH decreases, other precursor phases such as dicalcium phosphate dihydrate (DCPD) may form [26,28]. Therefore, it has been accepted that other calcium phosphate phases could actively participate in the crystallization reaction of biological (biogenic) apatites.

Hydroxyapatite powders can be synthesized with a range of production methods, however, one of the most commonly used method [32-34] is from an aqueous solution of $Ca(NO_3)_2$ and NaH_2PO_4. In this method, after filtering and drying, the product was calcined for about 3 hours at 900 °C to promote crystallization. Upon cold-press forming, the desired shape can be obtained and sintered for about 1 hour at about 1230°C to obtain a full densification. Upon sintering above 1250°C, hydroxyapatite shows a second-phase precipitation along grain boundaries and at multiple-grain junctions with formation of grain-boundary microcracks, with significant degradation of the mechanical properties. A range of alternative synthesis methods is covered in a number of reviews [26,28]. Pure synthetic HA, $Ca_{10}(PO_4)_6(OH)_2$, can be prepared by solid-state reaction, by hydrothermal or microwave method or by sintering apatite obtained from solution sol-gel methods [34-41]. Commercial HA is prepared by sintering apatite obtained by precipitation under basic conditions [42-43]. Unsintered apatite can be prepared by precipitation or by hydrolysis of DCPD or DCPA [28]. Coralline HA is prepared by hydrothermal hydrolysis of coral ($CaCO_3$ in calcite form) in the presence of ammonium phosphate [44,45]. Apatites derived from bovine bone are obtained by removing the organic phase without sintering or with sintering above 1000°C [46,47]. It may be expected that these 'HA' of different origin and preparation would have different crystallinity, composition and dissolution properties [27-28].

Substituted apatites recommended for bone graft or bone substitute biomaterial or as coating on dental and orthopedic implants include: carbonate apatite $Ca_{10}(PO_4)_6CO_3$ (CA); carbonate hydroxyapatite $(Ca,Na)_{10}(PO_4,CO_3)_6(OH)_2$ (CHA); Fluorapatite, $Ca_{10}(PO_4)_6(F,OH)_2$ (FA); strontium-substituted apatite, $(Ca,Sr)_{10}(PO_4)_6(OH)_2$ [48-49]. Substitutions in the apatite structure cause changes in its crystallographic, physical and chemical properties [50-51,27].

4. Coralline Apatites

Both natural coral and converted coralline hydroxyapatite have been used as bone grafts and orbital implants since the 1980s, as the porous nature of the structure allows in-growth of blood vessels to supply blood for bone, which eventually infiltrates the implant. Coralline apatites can be derived from sea coral. Coral is composed of calcium carbonate in the form of aragonite, is a naturally occurring structure and has optimal strength and structural characteristics. The pore structure of coralline calcium phosphate produced by certain species is similar to human cancellous bone, making it a suitable material for bone graft applications (Figure 1).

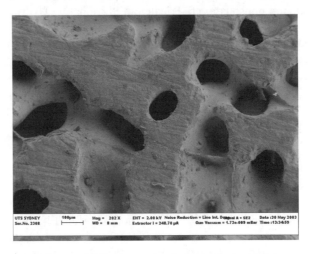

Figure 1. SEM of the coralline structure showing macro pores and Interpore regions.

Bone pore sizes range from 200 to 400 μm in trabecular bone and 1 to 100 μm in normal cortical bone and the pores are interconnected [52-54]. Porosity (macroporosity) is introduced in synthetic calcium phosphates (HA, β-TCP, BCP) by the addition of volatile compounds (for example, naphthalene or hydrogen peroxide) [35,43,53]. The original macroporosity of the coral is retained after the hydrothermal conversion to coralline HA [40]. The macroporosity of bovine-bone derived apatite is preserved from the original macroporosity of the bone.

In synthetic and natural bone graft materials the pore size and their interconnectivity are of utmost importance when hard and soft tissue in-growth is required. Kühne *et al.*, showed [54] that implants with average pore sizes of around 260 μm had the most successful in-growth as compared to no implants (simply leaving the segment empty). It was further reported that the interaction of the primary osteons between the pores via the interconnections allows propagation of osteoblasts.

Roy and Linnehan [41] were the first to use the hydrothermal method for hydroxyapatite formation directly from corals. It was reported that complete replacement of aragonite ($CaCO_3$) by phosphatic material was achieved at 260 °C and 103 MPa by using the hydrothermal process. During the hydrothermal treatment, hydroxyapatite replaces the aragonite whilst preserving the porous structure. The following exchange takes place:

$$10CaCO_3 + 6(NH_4)_2HPO_4 + 2H_2O \rightarrow Ca_{10}(PO_4)_6(OH)_2 + 6(NH_4)_2CO_3 + 4H_3CO_3 \quad (1)$$

The resulting material is known as coralline hydroxyapatite.

In 1996, HAp derived from Indian coral using the hydrothermal process was reported [55]. However, the resultant material was in powder form and required further forming and sintering. Aragonite to carbonate hydroxyapatite was achieved by using the microwave processing technique. Greater extents of conversion were reported [38]. Recently Hu et al., [37], and Ben-nissan et al. [56], succeeded in converting a high-strength Australian coral to monophasic hydroxyapatite by using a two-stage process in which the hydrothermal method was followed by a patented hydroxyapatite sol-gel coating process based on alkoxide chemistry. They reported a two-fold increase in the biaxial strength of the double-treated coral in comparison to the converted-only coral.

5. Calcium Phosphate Coatings

Due to its high porosity level (45-65%), porous hydroxyapatite has low mechanical strength and cannot be used under load-bearing conditions, unless fixed by external and internal support. For this reason hydroxyapatite has been applied as thin film coatings on metallic alloys. Of the metallic alloys investigated, titanium-based alloys have been shown to be the material of preference for thin film coatings [57]. Titanium alloys possess good mechanical strength and fatigue resistance under load-bearing conditions. They are lightweight, with high strength to weight ratios.

Of the coating techniques utilized, thermal spraying (plasma, and, to a lesser extent, flame spraying) tends to be the most commonly used and analysed. This technique has been faced with the challenge of producing a controllable resorption response in clinical situations due to a number of dissociation products (amorphous calcium phosphates, β-TCP and CaO) which have varying solubilities within the physiological environment. Besides the setbacks, thermally sprayed coatings are continually being improved by using different compositions and post-heat treatments which convert amorphous phases to crystalline calcium phosphates. Plasma coating of the macro-textured orthopedic implants is also used commercially, and other techniques involving less soluble fluorapatite compositions are also being investigated.

Other techniques that are capable of producing thin coatings include pulsed-laser deposition [58] and sputtering [59], which, like thermal spraying, involves high-temperature processing. Other techniques such as electrodeposition [60,61], and sol-gel [62] utilize lower temperatures and avoid the challenge associated with the structural instability of hydroxyapatite at elevated temperatures.

The advantages of the sol-gel technique are numerous; it results in a stoichiometric, homogeneous and pure coating due to mixing on the molecular scale; reduced firing temperatures due to small particle sizes with high surface areas; it has the ability to produce uniform fine-grained structures; the use of different chemical routes (alkoxide or aqueous-based); and their ease of application to complex shapes with a range of coating techniques: dip, spin, and spray coating. The lower cost due to the smaller amounts used and the low processing temperature has another advantage; it avoids the phase transition ($\sim$883°C) observed in titanium-based alloys used for biomedical devices.

6. Bone Graft Ceramics

Bone grafting is commonly used in orthopedic and maxillofacial surgery for non-union, the treatment of bridging diaphyseal defects, filling metaphyseal defects and mandibular reconstruction. An autogeneous bone graft (natural bone from the patient) is osteogenic (forms bone, due to living cells such as osteocytes or osteoblasts), osteoconductive (has no capacity to induce or form bone but provides an inert scaffold upon which osseous tissue can regenerate bone), and osteoinductive (stimulates cells to undergo phenotypic conversion to osteoprogenitor cells capable of formation of bone). Although there are no substitutes for autogenous bone, there are, however, synthetic alternatives.

Allografts such as demineralized bone (DBM) have been used as an alternative, but they have low or no osteogenicity, increased immunogenicity and resorb more rapidly than autogenous bone. In clinical practice, fresh allografts are rarely used because of immune response and the risk of transmission of disease. The frozen and freeze-dried types are osteoconductive but are considered, at best, to be only weakly osteoinductive. Freeze-drying diminishes the structural strength of the allograft and renders it unsuitable for use in situations in which structural support is required. Allograft bone is a useful material in patients who require bone grafting of a non-union but have inadequate autograft bone. Bulk allografts can be utilized for the treatment of segmental bone defects [63]. Calcium Sulphate (Plaster of Paris) or its composites is one of the oldest osteoconductive materials available. It has been used to fill bony defects, however, its main drawback is the chemical reaction that occurs during setting, which results in non-homogeneous structure with anisotropic properties. Demineralised bone matrix (DBM) was first observed by Urist in 1965 to induce heterotopic bone [64]. The active components of DBM are a series of glycoproteins, which belong to a group of transforming growth factor families (TGF-β) . The members of this group are responsible for the morphogenic events involved in the development of tissue and organs. Urist also isolated a protein from the bone matrix, which was termed the bone morphogenic protein (BMP) [64]. DBM is commercially available and used in the management of non-union of fractures. It is not suitable where structural support is required. To date, the main delay in developing clinical products has been the need to find a suitable carrier to deliver the BMP to the site at which its action is required. New-generation ceramic composites/hybrids could fill this gap. Experimentally, BMP-2 and OP-1® (BMP-7) have been shown to stimulate the formation of new bone in diaphysical defects in the rat, rabbit, dog, sheep and non-human primates [65]. BMPs and collagen with new calcium phosphate derivatives or composites could be used for bone remodeling where bone regeneration is required, such as therapeutic applications in osteoporosis.

The tissue engineering has been directed to take advantage of the combined use of living cells and tri-dimensional ceramic scaffolds to deliver vital cells to the damaged site of the patient. Feasible and productive strategies have been aimed at combining a relatively traditional approach such as bioceramics implants with the acquired knowledge applied to the field of cell growth and differentiation of osteogenic cells. A stem cell is a cell from the embryo, fetus, or adult that has the ability to reproduce for long periods. It also can give rise to specialized cells that make up the tissues and organs of the body. An adult stem cell is an undifferentiated (unspecialized) cell that occurs in a differentiated (specialized) tissue, renews itself, and becomes specialized to yield all of the specialized cell types of the tissue from which it originated. Cultured bone marrow cells can be regarded as a mesenchymal precursor cell population derived from adult cells. They can differentiate into different lineages: osteoblasts, chondrocytes, adipocytes and myocytes, and undergo limited mitotic divisions without expressing telomerase activity. When implanted onto immunodeficient mice, these cells can combine with mineralized tri-dimensional scaffolds to form a highly vascularized bone tissue. Cultured-cells/bioceramic composites can be used to treat full-thickness gaps of bone diaphysis with excellent integration of the ceramic scaffold with bone, and good functional recovery. Excellent innovative work is in progress and clinical applications are becoming quite common.

Bovine collagen mixed with hydroxyapatite is marketed as a bone-graft substitute, which can be combined with bone marrow aspirated from the site of the fracture. Although no transmission of disease has been recorded, its use will continue to be a source of concern. This material is osteogenic, osteoinductive and osteoconductive, however, it lacks the structural strength required.

7. Bone Cement Composites

There is a range of new generation bone cements; typically they are prepared like acrylic cements and contain a range of powders such as monocalcium phosphate, tri-calcium phosphate and calcium carbonate, which is mixed in a solution of sodium phosphate in various proportions. These cements are produced without polymerization and the reaction is nearly non-exothermic. The final compounds are reported to have strength of 10-100 MPa in compression, with 1-10 MPa in tension, although being very weak under shear forces. These composites are currently used in orthopedics in the management of fractures. Injection of calcium phosphate cement has been shown to be feasible and it does improve their compressive strength [66].

Preparation of hydroxyapatite/ceramic composites through the addition of various ceramic reinforcements has been attempted, with metal fibers [67], Si_3N_4 or hydroxyapatite whiskers [68], Al_2O_3 platelets [69] and ZrO_2 particles [67]. In many cases, the composites could not be successfully prepared due to the problems related to their poor densification.

Hydroxyapatite/metal and hydroxyapatite/polymer composites are two typical classes of materials, which have been examined for improving the toughness characteristics of synthetic hydroxyapatite. In both cases, a toughness improvement can be found, due to a crack-face bridging mechanism operated upon plastic stretching of metallic or polymeric ligaments. Zhang et al. [69] proposed a toughened composite consisting of hydroxyapatite dispersed with silver particles. This material was obtained by a conventional sintering method. It was reported that the toughness of these composites increased to 2.45 MPa $m^{1/2}$ upon loading the mixture, with (30 vol%) silver. The use of silver is not only for taking advantage of the ductility of silver in terms of fracture toughness, but also because silver is inert and has anti-bacterial properties [69].

8. Biomimetic Hybrid Composites

As stated earlier, bone is a composite in which nanosized apatite platelets are deposited on organic collagen fibers. If three-dimensional synthetic organic fibers can be fabricated into a composite structure, and then modified with functional groups, synthetic apatites, which are morphologically similar to the bone could be prepared. The resultant composite could be expected to exhibit bioactivity as well as mechanical properties analogous to those of the living bone. Various routes–some very novel–have been under investigation.

The conventional way to synthesize an inorganic material-based composite is to subject one of the constituents of the mixture to a specifically designed heat treatment. This process is also common in the biomaterials production arena; however, it is conceptually far from the biomineralization process, which occurs in nature. The natural process produces fine hybrid structures, which are hardly reproducible by classic consolidation processes. The traditional sintering route is not directly applicable to produce ceramic/polymer composites because no polymer will stand at the densification temperature of any ceramic material. Hydroxyapatite/polyethylene (HAp/PE) composites have been obtained by loading the polymeric matrix with the inorganic filler. In recent years, several research groups have demonstrated the feasibility of in vitro techniques for the synthesis of biomimetic material structures [42,70-75]. Currently a range of HAp/PE-based composites are produced and marketed by a UK-based company. The sophistication of the biomimetic route has not been paired yet and these techniques, so far, have not proved to be completely suitable for clinical applications [72]. It can be easily predicted that more and more dense bioceramic-based hybrid materials will be introduced, opening a completely new perspective on biomaterials production and application methods.

A new alternative route – based on an in situ polymerization process, carried out on an inorganic scaffold (with submicrometer-sized open porosity) – has also been proposed [69]. This is an intermediate method between the conventional sintering and biomineralization *in vitro*, because it still employs sintering for the preparation of the inorganic scaffold, however, the subsequent hybridization of the scaffold with the organic phase is carried out through a chemical route. This method, while aiming at relatively complex structural designs, enables the synthesis of biomimetic (hybrid) inorganic/organic composites through rather simple and easily reproducible methods.

A common characteristic of natural biomaterials such as bone, nacre, sea urchin tooth and other tough hybrid materials in nature is the strong microscopic interaction between the inorganic and organic phases. This characteristic allows the organic phase to act as a plastic energy-dissipating network, forming stretching (bridging) ligaments across the faces of a propagating crack on a nanoscale level. Such complexity has led to the common perception that, to mimic the natural designs, *in situ* synthesis techniques should be adopted. Precipitation of calcium carbonate or hydroxyapatite into a polymeric matrix, for example, has been proposed as a novel synthetic route to biomimetic composites [71-79]. Despite significant advances in understanding biological mineralization and developing new fabrication processes, the composites to date obtained by these methods are still too much within the embryonic stage for actual biomedical applications, due to their low structural performance.

The results of a study involving fracture analysis on two natural biomaterials, bovine femur and Japanese nacre (*Crassostrea Nippona*), in comparison with a synthetic hydroxyapatite/nylon-6 composite (obtained by *in situ* polymerization of caprolactam infiltrated into a porous apatite scaffold) showed that the high work of fracture achieved is about two orders of magnitude higher than that of monolithic hydroxyapatite, and it is due to stretching of protein or polymeric ligaments across the crack faces during fracture propagation [70].

Although the nanoscale modeling of synthetically manufactured hybrids and composites is still in its infancy, mimicking natural microstructures while using strong synthetic molecules may lead to a new generation of biomaterials, whose toughness characteristics will be comparable with the materials available in nature. A formidable challenge remains regarding the optimization of their morphology and bioactivity in these novel hybrid composites.

9. Concluding Comments

There are a number of technological problems with the synthesis and the use of current synthetic biomaterials. These can be attributed to the raw materials used, material purity, synthesis, production methods and related problems (defects), the lack of adequate information on the tissue-material interactions, surgical techniques and thepatients' well-being that influence the success rate of these implants.

Various methods and future alternatives have been proposed to solve these problems. The first is to solve the current problems (in the orthopedic area) with new-generation materials-based on biomimetics and/or nanotechnology. The second is related to tissue engineering, utilizing bioceramic or natural scaffolds filled with a range of biogenic materials (to allow and accelerate for example, cartilage regeneration or bone growth). Some of these new methods are promising and some are in the development stages. The third issue is genetic engineering and gene therapy (can we manipulate or trick cells?). This is a very novel, promising and unique approach, but possibly we have a long way to go.

Should we ignore the currently growing orthopedic and medical problems – mainly due to the ageing population – and develop, within 10-15 years, better solutions utilizing the second or the third approach, or go back and modify the currently existing and inadequate implants, while preparing for the future with the new methods?
The future and the patients' pain-free life and longevity might depend on the appropriate decisions being made by the current younger generation of scientists.

References

1. *The World Orthopaedic Market 2000–2001* (2002) Dorland's Biomedical/Knowledge Enterprises, Philadelphia.
2. Doremus, R.H. (1992) Review-Bioceramics, *J. Mater. Sci.* **27**, 285-297.
3. Boretos, J.W. (1987) Advances in Bioceramics, *Adv. Ceram. Mater.* **2**, 15-24.
4. LeGeros, R.Z. (1988) Calcium Phosphate Materials in Restorative Dentistry: A Review, *Adv. Den. Res.* **2**, 164-183.
5. Hench, L.L. (1991) Molecular Design of Bioactive Glasses and Ceramics for Implants, in W. Soga and A. Kato (eds.), *Ceramics: Towards the 21st Century*, Ceram. Soc. of Japan, pp. 519-534.
6. H. Aoki (1978) CaO-P$_2$O$_5$ Apatite. Japanese Patent JP 78110999.
7. Jarcho, M., Bolen, C.H., Thomas, M.B., Bobick, J.F. Kay, J.F., and Doremus, R.H. (1976) Hydroxyapatite synthesis and characterization in dense polycrystalline form, *J. Mater. Sci.* **11**, 2027-2035.
8. Kokubo, T., Shigematsu, M., Nagashima, Y., Tashiro, M., Nakamura, T., Yamamuro, T., and Higashi, S. (1982) Apatite-Wollastonite Containing Glass-Ceramic for Prosthetic Application, *Bulletin of Institute for Chemical Research Kyoto University* **60**, 260-268.
9. Kokubo, T. (1991) Novel Biomaterials Derived from Glasses, in W. Soga and A. Kato (eds.), *Ceramics: Towards the 21st Century*, J. Ceramic. Soc. Japan, pp. 500-518.
10. Hench, L.L, Splinter, R.J, Allen, W.C., and Greenlee, T.K. (1972) Bonding Mechanisms at the Interface of Ceramic Prosthetic Materials, *J. Biomed. Mater. Res. Sym.* **2**, 117-141.
11. Lutton, P. and Ben-Nissan, B. (1997) The Status of Biomaterials for Orthopaedic and Dental Applications: Part II- Bioceramics in Orthopaedic and Dental applications, *Mater. Tech.* **12**(3-4), 107-111.
12. De Jong, W.F (1997) La Substance Material darts lesos, *Rec. Tav. Chim.* **45**, 415-448.
13. LeGeros, R.Z. and LeGeros,J.P. (2003) Calcium phosphate bioceramics: past, present and future, in B. Ben-Nissan, D. Sher, and W. Walsh (eds.), *Bioceramics 15* Trans Tech Publications, Uetikon-Zurich, pp.3-10.
14. McConnell, D.J. (1965) Crystal chemistry of hydroxyapatite,its reaction to bone mineral. *Arch. Oral Biol.* **10**, 42.
15. LeGeros, R.Z. (1991) *Calcium Phosphates in Oral Biology and Medicine. Monographs in Oral Sciences.* Vol. 15, S. Karger, Basel.
16. LeGeros, R.Z. (1981) Apatites in biological systems, *Prog. Crystal Growth Character* **4**, 1-45.
17. Rey, C., Renuygopalakrishnan, V., Collins, B., Glimcher M.J. (1991) Fourier transform infrared spectroscopic study of the carbonate ions in bone mineral during aging, *Calcif. Tissue Int.* **49**, 251-258.
18. Brown, W.E., Smith, J.P., Lehr, J.R., and Frazier, A.W. (1962) Crystallographic and chemical reactions between octacalcium phosphate and hydroxyapatite, *Nature* **196**, 1051-1055.

19. Boskey, A. and Posner, A. (1984) Structure and formation of bone mineral, in *Natural and Living Biomaterials*, CRC Press, Boca Raton, pp. 127.

20. Brown, W., Schroeder, L., and Ferris, J. (1979) Interlayering of crystalline octacalcium phosphate and hydroxylapatite, *J. Phys. Chem.* **83** (11), 1385-1388.

21. Milev, A., Kannangara, G.S.K., and Ben-Nissan, B. (2003) Morphological stability of plate-like carbonated hydroxiapatite, *Mater. Letters* **57**, 13-14.

22. LeGeros R.Z. and LeGeros J.P. (2001) dense hydroxiapatite, in L.L. Hench and J. Wilson (eds). *An Introduction to Bioceramics,* World Scientific, London.

23. Emerson, W. and Fisher, E. (1962) The infrared spectra of carbonate in calcified tissues, *Arch. Ora. Biol.* **7,** 671-683.

24. Bigi, A., Cojazzi, G., Panzavolta, S., Ripamonti, A., Roveri, N., Romanello, M., Suarez, K.N., and Moro, L. (1997) Chemical and structural characterization of the mineral phase from cortical and trabecular bone, *J. Inorg. BioChem.* **68** (1), 45-51.

25. Pasteris, J.D., Wopenka, B., Freeman, J.J., K. Rogers, E. Valsami-Jones, J.A.M. Van der Houwen, and Silva, M.J. (2004) Lack of OH- in nanocrystalline apatite as a function of degree of atomic order: implications for bone and biomaterials, *Biomaterials,* **25**, 229-238.

26. LeGeros, R.Z. and Daculsi, G. (1990) In vivo transformation of biphasic calcium phosphate ceramics:ultrastructural and physico-chemical characterization, in T. Yamamuro, L. L. Hench and J. Wilson-Hench (eds.), *Handbook of Bioactive Ceramics*, CRC Press, Boca Raton, pp. 17-28.

27. LeGeros, R.Z., Daculsi, G., Orly, I., and Gregoire, M. (1991) Substrate surface dissolution and interfacial biological mineralization, in J.E.D. Davies (eds.), *The Bone Biomaterial Interface,* University of Toronto Press, Toronto, pp. 76-83.

28. Ben-Nissan, B., Chai, C., and Evans, L. (1995) Crystallographic and Spectroscopic Characterisation and Morphology of Biogenic and Synthetic Apatites, in D.L. Wise, D.J. Trantolo, D.E. Altobelli, M.J. Yaszemski, J.D. Gresser and E.R. Schwartz (eds.), *Encyclopedic Handbook of Biomaterials and Bioengineering*, Vol. 1 Part B: Applications, Marcel Dekker Inc., New York, pp. 191-221.

29. Albee, F.H. and Morrison, H.F. (1920) Bone Graft Surgery, *Ann Surg.* **71,** 32-39 (reprinted in *Clin. Orthro.&Rel. Res.*, 1996, 324, 5-12).

30. Gatti, A.M., Zaffe, D., and Poli, G.P. (1990) Behaviour of Tricalcium Phosphate and Hydroxyapatite Granules in Sheep Bone Defects, *Biomaterials* **11**, 513-517.

31. Klein, C.P.A.T, Driessens A.A., and de Groot K. (1984) Relationship Between the Degradation Behaviour of Calcium Phosphate Ceramics and Their Physical Chemical Characteristics and Ultrastructural Geometry, *Biomaterials* **5**, 157-160.

32. Nery, E.B., Lynch, K.L., Hirthe, W.M., and Mueller, K.H. (1975) Bioceramic implants in surgicaly produced infrabony defects, *J. Periodontol.* **46**, 328-339.

33. LeGeros, R.Z., Lin, S., Rohanizadeh, R., Mijare, D., and LeGeros, J.P. (2003) Biphasic calcium phosphate bioceramics, preparation, properties and applications, *J. Mater. Sci: Mater. Med.* **14**(3), 201-209.

34. Jarcho M. (1981) Calcium phosphate ceramics as hard tissue prosthetics, *Clin. Orthop.* **157,** 259-278.

35. deGroot R.Z.(1983) *Bioceramics in Calcium Phosphate*, CRC Press, Boca Raton.

36. Aoki H., Kato, K., Ogiso, M., and Tabata, T. (1977) Studies on the application of apatite to dental materials, *J. Dent. Eng.* **18**, 86-89.

37. Hu, J., Russell, J.J., Ben-Nissan B. and Vago, R. (2001) Production and Analysis of Hydroxyapatite from Australian Corals via Hydrothermal Process, *J. Mater. Sci. Letters* **20,** 85-87.

38. Pena, J., LeGeros, R.Z, Rohanizadeh R., and LeGeros, J.P. (2001) CaCO$_3$/Ca-P Biphasic Materials Prepared by Microwave Processing of Natural Aragonite and Calcite, in S Giannini and A. Moroni, (eds.), *Key Engineering Materials 192-195, Bioceramics 13*, Trans Tech Publications, pp. 267-270.

39. LeGeros, R.Z., LeGeros, J.P., Daculsi, G., and Kijkowska, R. (1995) Calcium phosphate biomaterials: preparation, properties and biodegradation, in D.L. Wise, D.J. Trantolo, D.E. Altobelli (eds.), *Encyclopedic Handbook of Biomaterials and Bioengineering. Part A. Materials*, Marcel Dekker, New York, 1429.

40. Ben-Nissan, B. and Chai, C. (1995) Sol-Gel Derived Bioactive Hydroxyapatite Coatings, in R. Kossowsky and N. Kossovsky (eds.), *Advances in Materials Science and Implant Orthopaedic Surgery*, NATO ASI Series, Series E: Applied Sciences, Vol. **294**, Kluwer Academic Publishers, 265-275.

41. Roy, D.M., Linnehan, S.K., (1974) Hydroxyapatite Formed from Coral Skeletal Carbonate by Hydrothermal Exchange. *Nature* **247**, 220-222.

42. Kokubo, T., Kim, H-M., Kawashita, M., and Nakamura, T. (2000) Novel Ceramics for Biomedical Applications, *J. Aust. Ceram. Soc.* **36** (1) 37-46.

102

43. Denissen, H.W. and deGroot, K. (1979) Immediate dental root implants from synthetic dense calcium hydroxyapatite. *J. Prosthet. Dent.* **42** (5), 551-556.
44. Shors, E.C., White E.W., and Kopchok, V. (1989) Biocompatibility, osteoconduction and biodegradation of porous hydroxyapatite, tri-calcium phosphate, sintered hydroxyapatite and calcium carbonate in rabbit bone defects, *Mater. Res. Soc. Proc.* **110** 211-217.
45. Hu, J., Fraser, R., Russell, J.J., Vago, R., and Ben-Nissan, B. (2000) Australian coral as a biomaterial: characteristics, *J. Mater. Sci. Technol.* **16**(6), 591-595.
46. Stavropoulos, A., Kostopoulos, L., Nyengaard, J.R., and Karring, T. (2004) Fate of bone formed by guided tissue regeneration with or without grafting of Bio-Oss or Biogran, *J. Clin. Periodon.* **31**(1), 30-39.
47. Rogers, K.D. and Daniels, P. (2002) An X-ray diffraction study of the effects of heat treatment on bone mineral structure, *Biomaterials* **23**(12), 2577-2585.
48. Okazaki, M., Matsumoto, T., Taki, T., Taira, M., Takahashi, J., and LeGeros, R.Z. (1999), Biocompatibility of CO_3-Apatite preparations with solubility gradients, in H. Ohgushi, G.W. Hastings, T. Yoshikawa (eds.), *Bioceramics 12,* World Scientific Publishing Co. Pte.Ltd., Singapore, pp. 337-340.
49. LeGeros, R.Z. (1967) Crystallographic studies on the carbonate substitution in the apatite structure, Ph.D. Thesis, New York University.
50. LeGeros, R.Z., LeGeros, J.P., Trautz, O.R., and Shirra, W.P. (1971) Conversion of monetite $CaHPO_4$ to apatites effect of carbonate on the crystallinity and the morphology of the apatite crystallites, *Adv. X-ray Anal.* **14**, 57-66.
51. LeGeros, R.Z. and Tung M.S. (1983) Chemical stability of carbonate and fluorite containing apatites, *Caries. Res.* **17**, 419-429.
52. Klawitter, J.J. (1979) *Basic Investigation of Bone Growth in Porous Materials,* PhD Thesis, Clemson University, Clemson.
53. Hubbard, W. (1974) *Physiological CaP as Orthopedic Implant Material,* PhD Thesis, Marquette University, Milwaukee.
54. Kuhne, J.H., Bartl, R., Frisch, B., Hammer, C., Jannson, V., and Zimmer, M. (1994) Bone formation in coralline hydroxyapatite–effects of pore size studied in rabbits, *Acta Orthopedica Scandinava* **65** (3), 246-252.
55. Sivakumar, M., Kumar, T.S., Shantra, K.L., and Rao, K.P. (1996) Development of Hydroxyapatite Derived from Indian Coral, *Biomaterials* **17**, 1709-1714.
56. Ben-Nissan, B., Russell, J.J., Hu, J., Milev, A., Green D. D., Vago, R., Walsh W., and Conway, R.M. (2000) Comparison of Surface Morphology in Sol-Gel Treated Coralline Hydroxyapatite Structures for Implant Purposes", *Bioceramics 13,* Trans Tech Publications, Key Engineering Materials, Vol. 192-195, 959-962.
57. Chai, C.S. and Ben-Nissan, B. (1993), Interfacial Reactions Between Hydroxyapatite and Titanium, *J. Aust. Ceram. Soc.* **29**, 81-90.
58. Cottel, C.M., Chrisey, D.B., Grabowski, K.S., Sprague, J.A., and Rossett, C.R. (1992) Pulsed Laser Deposition of Hydroxyapatite Thin Films on Ti-6Al-4V, *J. Appl. Biomater.* **3**, 87-93.
59. Ong, J.L., Lucas, L.C., Lacefield, W.R., and Rigney, E.D. (1992) Structure, Solubility and Bond Strength of Thin Calcium Phosphate Coatings Produced by Ion Beam Sputter Deposition, *Biomaterials* **13**(4), 249-254.
60. Ducheyne, P., van Raemdock, W., Heughebaert, J.C. and Heughebaert, M. (1986) Structural Analysis of Hydroxyapatite Coatings on Titanium, *Biomaterials* **7**, 97-103.
61. Zhitomirsky, I. and Gal-Or, L. (1997) Electrophoretic Deposition of Hydroxyapatite, *J. Biomed. Mater. Res.* **21**, 1375-1381.
62. Ben-Nissan, B., Green, D.D., Kannangara, G.S.K., and Milev, A. (2001) ^{31}P NMR Studies of Diethyl Phosphite Derived Nanocrystalline Hydroxyapatite, *J. Sol-Gel Sci. Tech.* **21**, 27-37.
63. Friedlaender, G.E., Strong, D.M., Tomford W.W., and Mankin, H.J. (1999) Long Term Follow-up of Patients with Osteochondral Allografts: A Correlation Between Immunogenic Responses and Clinical Outcome, *Orthop. Clin. North. Am.* **30**, 583-585.
64. Urist, M.R. (1965) Bone: Formation by Autoinduction, *Science* **12**, 893-899.
65. Cook, S.D., Wolfe, M.W., Salkeld, S.L., and Rueger, D.C. (1996) Effect of Recombinant human Osteogenic Protein-1 on healing of segmental defects in non-human primates, *J. Bone Joint Surg.* **77-A**, 734-750.
66. Kuhn, K-D. (2000) *Bone Cements*, Springer, NY.
67. Ruys, A.J., Zeigler, K.A., Brandwood, A., Milthorpe, B.K., Morrey, S., and Sorrell C.C. (1991) Reinforcement of Hydroxylapatite with Ceramic and Metal Fibres, in W. Bonfield, G.W. Hastings and K.E. Tanner (eds.), *Bioceramics Vol. 4*, Butterworth-Heinemann Ltd, London, pp. 281-286.

68. Ioku, K., Noma, T., Ishizawa, N., and Yoshimura, M. (1990) Hydrothermal Synthesis and Sintering of Hydroxyapatite Powders Dispersed with Si_3N_4 Whiskers, *J. Ceram. Soc. Japan Int. Ed.* **98**, 48-53.
69. Zhang, X., Gubbels, G.H.M., Terpstra, R.A., Metselaar, R. (1997) Toughening of Calcium Hydroxyapatite with Silver Particles, *J Mater Sci* **32**, 235-243.
70. Pezzotti, G. and Asmus, S.M.F. (2001) Fracture Behavior of Hydroxyapatite/Polymer Interpenetrating Network Composites Prepared by In Situ Polymerization Process, *Mater. Sci. Eng.* **316**, 231-237.
71. Sarikaya, M., Liu, J., and Aksay, I.A. (1995) Nacre: Properties, Crystallography, and Formation, in M. Sarikaya and I. A. Aksay (eds.), *Biomimetics: Design and Processing of Materials*, American Institute of Physics, NY, pp. 34-90.
72. Gautier, S., Champion, E., and Bernache-Assollant, D. (1997) Effect of Processing on the Characteristics of a 20vol% Al_2O_3 Platelet Reinforced Hydroxyapatite Composite, in L. Sedel and C. Rey (eds.), *Bioceramics, Vol .10*, University Press, Cambridge, pp. 549-552.
73. Bonfield, W., Grynpas, M.D., Tully, A.E., Bowman, J., and Abram, J. (1981) Hydroxyapatite Reinforced Polyethylene: A Mechanically Compatible Implant Material for Bone Replacement. *Biomaterials* **2**, 185-86.
74. Reis R.L., Cunha A.M., and Bevis M.J. (1997) Load-Bearing and Ductile Hydroxylapatite Polyethylene Composites for Bone Replacement, in L. Sedel and C. Rey (eds.), *Bioceramics 10*, University Press, Cambridge, pp. 515-518.
75. Almqvist, N., Thomson, N.H., Smith, B.L., Stucky, G.D., Morse, D.E., and Hansma, P.K. (1999) Methods for Fabricating and Characterizing a New Generation of Biomimetic Materials, *Mater. Sci. Eng. C* **7**, 37-43.
76. Calvert, P. and Mann, S. (1988) Review: Synthetic and Biological Composites Formed by In-Situ Precipitation, *J. Mater. Sci.* **23**, 3801-3806.
77. Ladizesky, N.H., Ward, I., and Bonfield, W. (1997) Hydrostatic Extrusion of Polyethylene Filled with Hydroxyapatite, *Polymers Adv. Tech.* **8**, 496-504.
78. Bonfield, W., Grynpas, M.D., and Bowman, J. (1980) UK Patent No. 2085461B.
79. Kim, T. (1998) *Design and development of a system to prepare and evaluate in vitro a PLGA/β-TCP composite"*, MSc Thesis. New York University College of Dentistry.

NEW BIOMIMETIC COATING TECHNOLOGIES AND INCORPORATION OF BIOACTIVE AGENTS AND PROTEINS

P. HABIBOVIC, F. BARRÈRE and K. DE GROOT
iBME, Twente University
P.O. Box 217, 7500 AE Enschede, The Netherlands

1. Introduction

The word biomimetics originates from Greek "Bios" (life, nature) and "Mimesis" (imitation, copy) and can be defined as "the investigation of the structures and functions of biological materials that allows possible future design and synthesis of engineered composites based on the principles obtained from the biological materials" as cited in [1].

The increase of life expectancy goes along with partial or full degradation of tissues and organs. The needs and advances in repair are therefore in continuous expansion. One of the examples in which a continuous development is visible is bone repair. Natural material sources are mainly utilized for bone repair. The source of these materials can be found in patient himself (autografting), in a donor (allografting) or in animals (xenografting). Autografts are the most favorable repairing materials. However, solely small volumes can be transferred from the donor site to the defect, and two surgical procedures are required. Allografting and xenografting are restricted by facts of limited supply, potential of disease transmission and host rejection. In addition, grafting can alter the initial properties of the implant, decreasing the quality of a graft. In order to overcome disadvantages of different kinds of grafts, new materials are continuously being synthesized by learning from nature. These materials are known as biomaterials.

Bone repairing material must be both biofunctional and biocompatible. Material biofunctionality concerns the ability of the implant to perform well for the purpose for which it was designed. Requirements for a suitable bone substitute are:

(a) Mechanical, such as tensile strength, fracture toughness, elongation at fracture, fatigue strength, Young's modulus;
(b) Physical, such as density or thermal expansion;
(c) Chemical, such as degradation resistance, oxidation and corrosion [2].

Material biocompatibility can be defined as the biological acceptance of the implant. Osborn and Newesely [3] classified biocompatibility into three categories based on their reaction with the surrounding tissue. Biotolerant materials are characterized by a distant osteointegration with the formation of fibrous tissue layer between implant and bone.

R.L. Reis and S. Weiner (eds.),
Learning from Nature How to Design New Implantable Biomaterials, 105-121.

Bioinert materials are incapsulated with a thin fibrous layer. Finally, bioactive materials are able to enhance bonding with bone.

Various types of synthetic bone substitutes have been developed in order to comply with material biofunctionality and biocompatibility, such as metallic implants (titanium (Ti) and its alloys, stainless steel, cobalt-chromium alloys), ceramics (calcium-phosphate (CaP), aluminum oxide, glass ceramic) and polymers (silicon, rubber, (poly) methyl methacrylate, polylactide). Depending on the type of repair, and therefore the biological function that needs to be replaced, one or other material will be chosen. However, the complexity of the biological performance of bone makes it often necessary to combine different biomaterials in order to design a good bone substitute.

An example of such a combination is the application of CaP coatings on metallic implants. Thereby, high mechanical strength of metals is combined with the osteoconductive properties of CaPs, making such an implant well integrated with the host bone on e.g. load-bearing sites. Plasma-spraying (PS) is a well-known method of applying CaP coatings on metals. PS coatings, mainly hydroxyapatite (HA), on titanium alloy (Ti6Al4V) prostheses have widely been used in orthopaedic surgery to reconstruct hip and knee joints. Earlier investigations have shown that these coatings can successfully enhance clinical success to less than 2% failures after 10 years [4]. Despite excellent clinical performances, the plasma spray process is limited by intrinsic drawbacks. For instance, this line-of-sight process takes place at high temperatures. The process is, therefore, limited to thermally stable phases like HA, and the incorporation of growth factors that stimulate bone healing is impossible. Furthermore, this process cannot provide even coatings on porous metal surfaces.

Recently, other techniques have been studied to improve the quality of coatings, such as electrophoretic deposition [5], sputter deposition [6], and sol-gel [7]. Nevertheless, the deposition of apatite coatings from simulated body fluids (SBFs) offers the most promising alternative to plasma spraying and other coating methods. The biomimetic approach has four main advantages: (a) it is a low temperature process applicable to any heat-sensitive substrate including polymers [8], (b) it forms bone-like apatite crystals having high bioactivity and good resorption characteristics [9], (c) it is evenly deposited on or even into porous or complex implant geometries [10] and (d) it can incorporate bone growth stimulating factors.

This biomimetic approach consists of soaking metal implants in simulated body fluids at physiological temperature and pH. Apatite coatings have successfully been formed by immersion of chemically pre-treated substrates such as glasses, metals and polymers in metastable SBFs [11-13]. Although SBF mimics the inorganic composition, the pH and the temperature of human blood plasma, it is unknown whether these conditions are optimal for a coating process. Indeed, a thin apatite layer has previously been obtained on pre-treated substrates by using long immersion time (i.e. 7-14 days) with daily refreshment of SBFs [14-16]. The difficulty results from the metastability of SBF. The process requires replenishment and a constant pH to maintain supersaturation for apatite crystal growth. As a result of the low solubility product of HA and the limited concentration range for the metastable phase, this operation is extremely difficult and might lead to local precipitation or uneven coatings. Such an intricate and long process can hardly be applicable in the coating prostheses industry.

We have developed a new biomimetic route for coating metallic implants with uniform and thick carbonated apatite (CA) and octacalcium phosphate (OCP) layers. In this chapter, biological performances of these novel coatings are described, as well as the possibility to incorporate bioactive agents and relevant proteins into the coatings in order to further improve these performances.

2. Preparation of Biomimetic Carbonated Apatite (CA) and Octacalcium Phosphate (OCP) Coatings

2.1. CARBONATED APATITE

2.1.1. *Method*

As earlier described [17] the process of producing biomimetic CA layer on Ti alloy (Ti6Al4V) surface consists of two steps. In the first step, the heterogeneous nucleation of a thin and amorphous Ca-P layer on the metal surface is obtained. During the second step, the growth of a thick and crystallized apatite coating on the implants is favored due to lower Mg^{2+} and HCO_3^- contents.

SBF-A and the SBF-B solutions (table 1) are prepared according to Kokubo's SBF solution [18], excluding TRIS-buffer, K^+- and SO_4^{2-} ions. The first-step solution, SBF-A is 5 times more concentrated in NaCl, $MgCl_2.6H_2O$, $CaCl_2.2H_2O$, $Na_2HPO_4.2H_2O$ and $NaHCO_3$ than Kokubo's SBF solution. SBF-B solution has the same composition as SBF-A solution, but the contents of so-called inhibitors of crystal growth (i.e. Mg^{2+} and HCO_3^-) are lower (table 1). All salts are precisely weighed (±0.01 g), and dissolved in demineralized water under supply of CO_2 gas at a flow of 650 l/min for 20 minutes and stirring at a speed of 250 rounds per minute (rpm).

TABLE 1. Inorganic composition of Human Blood Plasma (HBP), Simulated Body Fluid (SBF) and solutions SBF-A and SBF-B

	Ion concentration (mM)							
	Na^+	K^+	Ca^{2+}	Mg^{2+}	Cl^-	HPO_4^{2-}	HCO_3^-	SO_4^{2-}
HBP	142.0	5.0	2.5	1.5	103.0	1.0	27.0	0.5
SBF	142.0	5.0	2.5	1.5	148.8	1.0	4.2	0.5
SBF-A	714.8	-	12.5	7.5	723.8	5.0	21.0	-
SBF-B	704.2	-	12.5	1.5	711.8	5.0	10.5	-

The process is performed in a 7L bioreactor (Applikon Dependable Instruments, Schiedam, The Netherlands) (Figure 1).

The implants are first soaked in an SBF-A solution for 24 to seed the metal surface with CaP nuclei. During this process, the temperature is kept at 37°C and the solution is stirred at 250 rpm. In order to exchange CO_2 gas from the solution with air, a top aeration at a flow of 450 ml/min is performed. In addition, air is added through the solution (flow 500 ml/min) after the pH had reached a value of 7.1. At the end of the process the coated samples are cleaned with demineralized water and dried in air overnight. Then, the implants are soaked for another 24 hours in the SBF-B solution under crystal growth conditions. Temperature is maintained at 50°C and the stirring speed at 250 rpm. The same aeration system is used as described above.

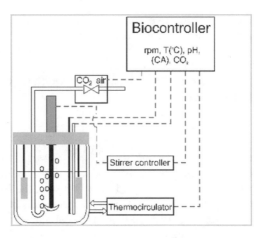

Figure 1. Bioreactor for biomimetic CA coating. After Habibovic *et. al.* [17].

Coated implants are investigated macroscopically and microscopically using an environmental scanning electronic microscope with field emission gun (ESEM-FEG). The thickness of the coating is determined by using Eddy-current (electromagnetic) test method according to ASTM E376-96 [18]. Both coating and precipitate formed in the solution are investigated by Fourier transform infra-red spectroscopy (FTIR) and X-ray diffraction (XRD). The crystallinity of the coating is determined according to ASTM STP 1196 [19], and the Ca/P ratio according to NF S94-066 [20].

2.1.2. Coating Characterization

After 24 hours of soaking in SBF-A solution, the implants are covered with a Ca-P film with a thickness of approximately 3 μm. The ESEM photos show that this Ca-P layer is uniformly deposited on the Ti6Al4V surface (figure 2a). The coating is dense and consists of globules. This thin coating exhibits some cracks, probably formed during the drying process of the coated samples. The FTIR spectrum of the coating (figure 3a) showed featureless phosphate- and carbonate bands. Intense and broad bands assigned to O-H stretching and bonding are observed at, respectively, 3435 and 1642 cm^{-1}. Additionally, three bands at 868, 1432 and 1499 cm^{-1} corresponded to CO_3 groups. Finally, the one-component bands at 560 cm^{-1} and 1045 cm^{-1} show the presence of PO_4. This FTIR spectrum is characteristic of carbonated amorphous Ca-P. The XRD pattern of the SBF-A precipitate (Figure 4a) corroborates the FTIR results, showing only the halo characteristic of an amorphous phase.

The immersion process in the SBF-B solution is used to develop a second coating layer on the seeded Ti6Al4V surface resulting from the first coating process. After 24 hours of soaking, the coating reaches a thickness of ± 30 μm, homogeneously covering the total metal surface and consisting of well formed crystals with a size of 1-3 micrometers (Figure 2b).

The FTIR spectrum of the precipitate from the solution SBF-B, gathered between 400 and 4000 cm^{-1} (figure 3b), exhibits the characteristics of a carbonated apatite type A-B. The bands at 3435 and 1640 cm^{-1} correspond to O-H groups. CO_3 group bands are observed at 1493, 1417 and 872 cm^{-1} while bands at 1061, 599 and 563 cm^{-1} are assigned to PO_4/HPO_4 groups.

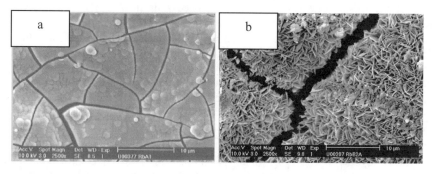

Figure 2. ESEM photographs (magnification 2500x) of (a) SBF-A coating and (b) SBF-B coating. After Habibovic *et. al.* [17].

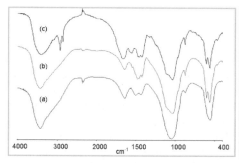

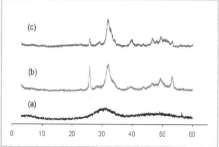

Figure 3. Infra-red spectrs of (a) SBF-A coating, (b) SBF-B coating and (c) bone. After Habibovic *et. al.* [17].

Figure 4. XRD patterns of (a) SBF-A coating, (b) SBF-B coating and (c) bone. After Habibovic *et. al.* [17].

The XRD pattern exhibits broad diffraction lines (figure 4b). The position and intensities of these diffraction lines indicate an apatitic structure. The peak at $2\theta = 32.1°$ corresponds to the overlapping of (211), (112), (300) and (202) diffraction peaks. Additionally, the peak at $2\theta = 25.8°$, corresponding to (002) diffraction plane, indicats that SBF-B coating consists of small apatitic crystals. The crystals have a size of 2-3 μm. The crystallinity of the coating is around 75% and its Ca/P ratio 1.67.

As shown in another study [21] the final apatite coating is well adhered to the Ti6Al4V substrate and no coating delamination or spalls are observed during the scratch test.

2.2. OCTACALCIUM PHOSPHATE

2.2.1. *Method*

Similarly to the method of preparation of CA described above, the process of OCP growth consists of two steps [22]. Like for the CA coating, first step is used for seeding the metal surface with CaP nuclei. The second step consists of immersion of the pre-seeded implants into Simulated Phisiological Solution (SCS) in order to grow the final

OCP coating. This process is taking place in closed vials at 37° C for 48 hours with one replenishment and under a continuous stirring at a speed of 100 rpm. Table 2 shows the composition of the two solutions. Final coating is fully characterized by using the same techniques as described above for the CA coating.

TABLE 2. Inorganic composition of Human Blood Plasma (HBP), Simulated Body Fluid (SBF) and solutions SBF-A and SCS

	Ion concentration (mM)							
	Na^+	K^+	Ca^{2+}	Mg^{2+}	Cl^-	HPO_4^{2-}	HCO_3^-	SO_4^{2-}
HBP	142.0	5.0	2.5	1.5	103.0	1.0	27.0	0.5
SBF	142.0	5.0	2.5	1.5	148.8	1.0	4.2	0.5
SBF-A	714.8	-	12.5	7.5	723.8	5.0	21.0	-
SCS	140.4	--	3.1	--	142.9	1.86	--	--

2.2.2. Coating Characterization

After 48 hours of soaking in SCS, surface of the metal is homogeneously covered with a 55 μm thick coating. As shown in the figure 5, the coating consists of large crystals (30-60 μm) that perpendicularly grow onto the metal surface.

Figure 6 shows the FTIR spectrum of the coating. This spectrum displays sharp P-O bands at 1110 and 1070 and 1023 cm^{-1} (P-O stretching in phosphate and hydrogenophosphate), and a HPO_4^{2-} band at 906cm^{-1} (P-O stretching in HPO_4^{2-}), typical of an OCP structure. The two sharp P-O bands at 560 cm^{-1} and 600 cm^{-1} (P-O deformation in PO_4^{3-}) exhibit small shoulders at 624 cm^{-1} (H$_2$O libation) and 526cm^{-1} (P-O twisting in HPO_4^{2-}) [23].

Figure 5. ESEM photograph (magnification 5000x)
of OCP coating. After Barrère *et. al.* [22].

XRD pattern of the OCP coating (figure 7) displays sharp diffraction lines at 2θ = 4.7° corresponding to (010) diffraction line and at 2θ = 25.5° corresponding to the (002) plan, both typical of triclinic OCP crystals [23]. Furthermore, the line at 2θ = 16.8° corresponds to the plan (-101), indicating that a highly crystalline coating has grown on Ti alloy surface. The crystallinity of the coating is around 100% and its Ca/P ratio 1.33.

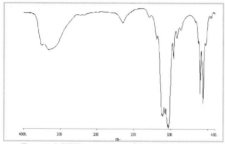 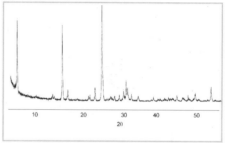

Figure 6. FTIR spectrum of OCP coating. After Barrère *et. al.* [22].

Figure 7. XRD pattern of OCP coating. After Barrère *et. al.* [22].

3. Mechanisms of Biomimetic Approach

Both above described processes are based upon the precipitation mechanism from a physiological solution. The process of CA coating formation shows that highly concentrated solutions can be obtained by addition of the mildly acidic gas, CO_2. It is, namely, well known that the solubility of Ca-P salts increases with the decrease of pH [24]. Dissolution of CO_2 gas results in a pH decrease due to the formation of carbonic acid H_2CO_3 (reaction (1)), after which the acid immediately dissociates in HCO_3^- and CO_3^{2-} species (reactions (2) and (3)).

$$CO_2 + H_2O \leftrightarrow H_2CO_3 \tag{1}$$
$$H_2CO_3 + H_2O \leftrightarrow HCO_3^- + H_3O^+ \tag{2}$$
$$HCO_3^- + H_2O \leftrightarrow CO_3^{2-} + H_3O^+ \tag{3}$$

When CO_2 gas gradually releases from the solutions, the pH slowly increases again (Figure 8). In the case of NaCl reference (NaCl in distilled water), the pH increases progressively from the start until the end of the process, slowly approaching the pH of water (6.5). The presence of HCO_3^-- and HPO_4^{2-} ions in SBF solutions, together with the CO_2 and HCO_3^- ions that are formed during the dissolution of CO_2 gas (reactions (2) and (3)), leads to the formation of buffered solution. This might explain the fact that the initial pH values of SBF-A and SBF-B solutions are higher than that of the NaCl reference as shown in the Figure 8. During the processes in SBF-B solution, a drop in the pH is observed simultaneously with the start of precipitation in the solution. This drop of the pH may be explained by the start of the precipitation of crystalline phase according to the reaction (4).

$$10Ca^{2+} + 6PO_4^{3-} + 2OH^- \leftrightarrow Ca_{10}(PO_4)_6(OH)_2 \tag{4}$$

Due to the decreased amount of OH^- ions in the solution, a pH drop is observed. After precipitation, CO_2 gas continues to release from the solution, and thereby, pH increases progressively until the end of the process.

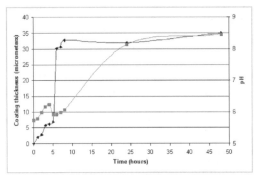

Figure 8. (•) pH and (♦) coating thickness versus
immersion time in SBF-B solution. After Habibovic *et. al.* [17].

In the SBF-A solution, more HCO_3^- species are present in comparison with the SBF-B solution (table 1). Therefore, the buffering capacity of CO_2/HCO_3^- couple is higher. This might explain a slightly higher initial pH and absence of pH drop during the process in SBF-A solution in comparison with the process in the SBF-B solution.

Another factor that may influence the pH evolution during the coating process is the amount of the NaCl species, i.e. ionic strength of the solution [25]. The rate of the CO_2 release from the solution is higher in case of low ionic strength, followed by a higher pH increase. This causes an early and sudden precipitation in the solution. Consequently, the supersaturation of the solution is markedly lowered and less ionic species are available in the solution for the Ca-P nucleation on the substrate. Furthermore, there is a relation between Ca-P concentration and the pH at which precipitation occurs: the lower the concentration, the higher the precipitation-pH [24].

According to the literature and our experience, a coating that successfully enhances osteo-integration of metal implants needs to be thick and crystallized enough to accommodate the bone healing process. Preliminary experiments were performed in order to grow a crystalline coating on metal surface without immersion in SBF-A solution. By immersion of cleaned titanium alloy directly in SBF-B solution, a loose and non-uniform layer was obtained, which shows the relevance of amorphous pre-coating. That is why our biomimetic costing process consists of two steps.

Immersion of the implants in SBF-A solution is necessary for seeding the metal surface with Ca-P nuclei. During this nucleation process, Ca-P seeds are precipitated in the solution and on the metal surface. Some of these nuclei can dissolve in the solution, and some of them can expand in their size. Homogeneous nucleation (precipitation) occurs spontaneously in the solution and can proceed by other seeds formed in the mean time. Heterogeneous nucleation, on the other hand, takes place on the metal surface. Both, homogenous and heterogeneous nucleation are in competition during the process in the SBF-B solution. However, nuclei are energetically more stable on the seeded metal surface than in the solution. It is, therefore, essential to provide the metal surface with a thin and uniform primer Ca-P layer for subsequent growth of the final coating. The kinetics of the process in SBF-A solution was reported in detail by Barrère *et.al.* [26].

After reaching their critical size, seeds can start growing into crystals. The nucleation and growth kinetics of the crystal depend on the temperature, pH, composition and saturation of the solution. Calcium and phosphate ions are responsible for the formation of the calcium phosphate layer on the metal surface, while magnesium and carbonate ions favor heterogeneous nucleation rather than crystal growth. As given in table 1, the SBF-B solution has the same composition as SBF-A solution, but the contents of Mg^{2+} and HCO_3^- ions are lower. Resulting from lower amounts of these so-called crystal growth inhibitors [27-34], by immersion in SBF-B solution a crystalline apatite phase is formed and a drop in the pH is observed at the start of precipitation. Furthermore, a lower amount of HCO_3^- ions, in comparison to SBF-A solution, decreases the buffering capacity of CO_2/HCO_3^- couple and thereby, variations in the pH can be observed. With a lower amount of Mg^{2+} in SBF-B solution, the Ca-P precipitation is accelerated and the growing coating becomes more crystallized. Earlier research [35-38] proved that crystal growth from a supersaturated solution occurs very fast. This can explain the fact that, within 1 hour, the coating thickness grows by 25 μm. Due to precipitation and crystal growth, the amount of Ca and P in the solution decreases. One hour after the start of precipitation, the contents of Ca and P are too low for further growth. Between the end point of crystal growth and the end of the process, there is an equilibrium between the amount of calcium and phosphate in the coating and in the solution. However, coating might dissolve and re-precipitate onto surface, resulting in a more homogeneous and denser coating at the end of the process.

Similarly to the first process, the process of OCP coating formation starts with the CaP seeded metal surface. The supersaturation of the SCS increases at the vicinity of the pre-seeded substrate and thereby leads to the heterogeneous nucleation of the CaP crystals. When the first CaP layer dissolves, it can initiate the deposition of OCP. This mechanism can be illustrated by the following reaction:

$$8Ca^{2+} + 6HPO_4^{2-} + 5H_2O \leftrightarrow Ca_8H_2(PO_4)_6, 5H_2O + 4H^+ \qquad (5)$$

Besides the influence of the seeding role of the CaP film on the nature of the subsequent CaP coating, authors have shown that the formed phase depends on the characteristics of the soaking supersaturated solution [37,39-41]. In the biomimetic process of OCP formation, SCS is supersaturated for both HA and OCP phase. HA is the most thermodynamically stable phase at physiological conditions, but OCP, thus more soluble than HA, is kinetically favored [42]. Subsequently, OCP, being a metastable phase, can transform into more thermodynamically stable CaP phases [37,43]. However, other studies have isolated stable OCP crystals at physiological conditions. In our approach, the thick OCP layer is stable while the immersion time increases. The OCP coating continuously grows while the supersaturation of the SCS decreases.

4. Biological Performances of Biomimetic Coatings

Biological performance in terms of osteointegration and osteoinduction of both CA and OCP biomimetic coating has been tested in different *in vivo* studies.

In the goat study by Barrère *et al.* [44], porous tantalum (Ta) and dense Ti alloy, coated with OCP coating were tested ectopically. Porous Ta and dense Ti alloy cylinders (∅5x10 mm) were implanted in back muscles of Dutch milk goats for 12 and 24 weeks. With regard to intramuscular implantation, OCP coating was shown to induce bone formation in extraskeletal sites. Both non-coated metals that were used as controls did not show any ectopic bone formation. On the contrary, porous structures coated with OCP coating induced bone formation. However, bone was not found on the flat coated surface of the dense Ti implant. These observations suggest that both, presence of a CaP coating and implant geometry play an important role in osteoinductive potential of an implant.

In the same study, CA- and OCP coated porous Ta cylinders (Ø5x10 mm) were inserted in femoral condyle of the goats using Teflon spacers for 12 and 24 weeks. Direct bone contact was noted for CA and OCP coated porous Ta, while a fibrous tissue layer was often detected between the non-coated implant and the newly formed bone. Therefore, both CaP coatings allowed a direct bonding between the implant and the host bone. In none of the implants, the gap of 1mm was completely healed, but on the contrary to CA coated implants, bone was found in the center of OCP coated implant. Differences could also be found in the *in vivo* degradation behavior of the two implants. After 12 weeks of implantation CA coating had completely disappeared, while the OCP coating partially remained. In a bony environment, the resorption of the coating strongly depends on osteoclastic activity, ruled by the physicochemistry of the CaP coating [45,46]. The more soluble CaP phase is, the less osteoclastic activity can be found. In the case of biomimetic CA and OCP, Leeuwenberg *et al.* [9] have shown that resorption pits further extend on CA coating than on OCP coating *in vitro*. Therefore it could be expected that the osteoclastic resorption *in vivo* could be higher in case of CA coating, which on its turn could negatively effect bone formation on CA coated implants. Besides the differences in physicochemistry, the initial microstructure of the two coatings differs as well: OCP exhibits a rough surface due to the sharp vertical crystals, while CA has a relatively smooth surface composed of globules. This initial microstructure can influence osteoinductive potential of CaP coatings: a rougher microstructure is beneficial for bone induction as compared to a smoother one. The relatively rough structure may also positively influence bone formation at orthotopic sites.

Similar to the previously described study, Habibovic *et al.* [47] looked into osteointegration of OCP coated porous Ti alloy implant in a goat model. Thereby, OCP coated and uncoated porous Ti alloy implants were inserted in femoral diaphysis of adult Dutch milk goats, for 6 and 12 weeks. Figure 9 shows the results of orthotopic implantation. These results show that the presence of OCP coating on a porous metal implant significantly increases its osteointegration after both 6 and 12 weeks of implantation.

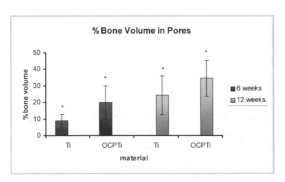

Figure 9. Histomorphometrical results of % bone volume formed in the pores of the implant after bony implantation. After Habibovic *et. al.* [47].

TABLE 3. Bone incidence after intramuscular implantation
in goats. After Habibovic *et. al.* [47]

Material	6 weeks	12 weeks
Ti6Al4V	0/10	0/10
OCP-Ti6Al4V	4/10	6/10
HA	0/10	0/10
OCP-HA	2/10	0/10
BCP	3/10	6/10
OCP-BCP	4/10	6/10
OCP-PEGT/PBT	0/10	0/10

Yuan *et al.* [48] looked into osteoinductive potential of OCP coated porous Ta implants in the muscles of adult dogs for 3 months. After explantation, abundant bone formation was found in all coated implants, while none of the uncoated implants induced ectopic bone.

As previously mentioned, the advantage of biomimetic coating method in comparison with others is that it is taking place at physiological conditions, making it applicable for different kinds of implants. In a study in NZW rabbits by Du *et al.* [49] with CA-coated and uncoated porous PEGT-PBT implants was shown that osteoconductive properties of the polymer are significantly improved by the CaP coating.

Another goat study [50] was used to investigate general effect of OCP coating on osteoinductive performance of various biomaterials. Different porous materials were coated with OCP coating, and uncoated implants were used as control. Materials were implanted in back muscle of adult goats for 6 and 12 weeks. As can be seen from table 3, biomimetic OCP coating is able of improving osteoinductive behavior of different kinds of orthopedic implants.

5. Incorporation of Bioactive Agents and Proteins into Biomimetic Coatings

Physiological conditions, at which biomimetic coating processes are taking place, make it possible to incorporate biologically active molecules such as osteogenetic agents and growth factors into the coatings. In addition, because the degradation of these coatings *in vivo* should result in a gradual exposure and release of incorporated molecules, they are of great potential value as drug-carrier system in orthopaedics.

5.1. INCORPORATION OF PROTEINS INTO BIOMIMETIC COATINGS

In the study by Liu *et al.* [51], it has been shown that large protein molecules like bovine serum albumin (BSA) can successfully be coprecipitated with the biomimetic OCP coating on Ti alloy implants, and that the final release of this protein takes place gradually. The process of protein incorporation in this study was performed by adding various concentrations of BSA into the SCS solution in which the second step of the OCP process is taking place. Anti-BSA immunological staining was used to demonstrate the incorporation of protein into the CaP coatings. The release of protein and calcium from coatings was analyzed in vitro by dissolving the coating in simulated physiologic solution (SPS) at pH5.0 and pH7.3 and measuring BSA and Ca-ions concentration at 2, 4, 6, 24 and 144 hours. Enzyme linked immunosorbent assay (ELISA) and bicinchoninic acid (BCA) assay measured the amounts of protein loaded and released.

The results of this study showed that the protein was successfully co-precipitated with CaP through the whole coating layer. ESEM micrographs of OCP coating with and without BSA showed different morphologies. Due to the protein incorporation, CaP crystals are becoming smaller and less sharp. With increasing concentrations of proteins in the calcifying solution, loading amounts of BSA in the coating were increased from 0.7μg/mg to 16μg/mg and the coating changed its structure from a highly crystalline octacalcium phosphate (OCP, $Ca_8H_2(PO_4)_6.5H_2O$) to a poorly crystallized calcium-deficient apatite ($Ca_{10-x} (PO_4)_{6-x} (HPO_4)_x (OH)_{2-x}$). The crystal size decreased and coatings become denser and thinner. During a 6-day release test, coating thickness decreased by around 20% and about 1% of totally incorporated protein was released (Figure 10). The release of Ca^{2+} in simulated physiological solution at pH 5 was much slower from the coating with incorporated protein as compared to coating produced without protein. One notices an initial rapid release (from loosely adsorbed BSA), followed by a slow release from BSA incorporated in the bulk crystals.

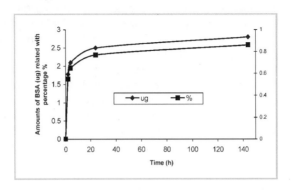

Figure 10. BSA release from biomimetic OCP coating in SPS pH5.0. After Liu *et. al.* [51].

It is believed that protein adsorption on Ca-P is electrostatic in nature and that several different functional groups of the BSA macromolecule may be involved into the adsorption process. Preferential binding sites for the protein on the surface are Ca^{2+} and PO_4^{3-} ions providing the major driving force for protein adsorption [52]. The highly ionic surface of Ca-P coating does not only attract the protein, but also exerts a greater electrostatic force on its functional groups. Beside adsorption, during the biomimetic process most BSA is incorporated into the CaP crystals. Thus, BSA is not only physically adsorbed onto the crystal plate surfaces, but also incorporated into the Ca-P crystal lattice also suggested by the fact that BSA significantly modifies the coating thickness as well as morphology and composition of Ca-P crystals.

Under physiologic situations, higher concentration of proteins on the surface will interfere with the re-precipitation of CaP, resulting in a lower mineralization rate. These results suggest that BSA might be an inhibitor of CaP crystal growth.

Although BSA is an elegant and easy model for testing protein incorporation potential, other proteins, such as bone morphogenetic proteins (BMPs) are more relevant for clinical purposes. In a study by Liu *et al.* [53] rh-BMP-2 was incorporated into OCP coating on Ti alloy implants, and subsequently implanted in a rat model to investigate protein release and biological performance.

Similar to the previously described study, rhBMP-2 (10µg/ml) was added to the simulated calcifying solution, in which the second step of the OCP coating process is taking place. The amount of rhBMP-2 incorporated into each coating was determined by ELISA. Ti alloy coated discs, with incorporated BMP-2 were then subcutaneously implanted in rats. Samples were retrieved at 7-day intervals over a period of 5 weeks.

Results of this study showed that all Ti alloy implants were uniformly coated with a 25-µm-thick layer containing 1.6 µg of rhBMP-2 at a concentration of 0.5 µg per mg of CaP. Coatings prepared in the absence and presence of rhBMP-2 had a similar morphological appearance, each being entirely comprised of plate-like crystals. FTIR revealed the calcium phosphate crystals to have a typical OCP structure. No structural change was elicited by the presence of BMP-2.

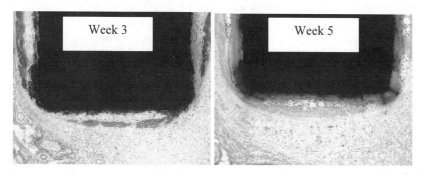

Figure 11. Ectopic bone formation in rats by rh-BMP-2 released from biomimetic OCP coating. After Liu *et. al.* [53].

One week after implantation, small numbers of foreign-body giant cells observed aligning the surfaces of BMP-2-containing coatings, but there was no evidence of osteogenic activity. This changed at the end of the second week and increased steadily thereafter. By the end of the fifth week, implants were completely encapsulated by woven bone (figure 11). At many locations, bone marrow impinged directly on the coatings, without intervening osseous tissue. Bone was deposited not only on the coated surfaces of the implants but also within the surrounding connective tissue. Ossification occurred by an intramembranous process; there was virtually no evidence of an enchondral mechanism having been followed. Parts of the coatings that were not covered with bone or bone marrow were occupied by foreign-body giant cells. Inflammatory cells were not encountered in great numbers around the implants. During the course of the 5-week follow-up period, the coatings underwent gradual degradation

In this study, BMP-2 was released from the calcium phosphate latticework at a sufficiently slow rate and at a sufficiently high level to support and sustain ectopic bone formation in a rat model. The finding that the biomimetic coating can act in this capacity is potentially of great clinical interest in orthopaedic and dental implant surgery. In previous studies, osteogenic activity at ectopic sites has been generally shown to follow an enchondral course. The observation that bone formation occurred predominantly via a direct (i.e., intramembranous) mechanism in such a situation signifies that it took place under conditions of high mechanical stability [54]. This circumstance bodes well for the therapeutic potential of our BMP-2-carrier system in clinical orthopedics and dentistry.

5.2. INCORPORATION OF ANTIBIOTICS INTO BIOMIMETIC COATINGS

Besides the incorporation of the agents that can stimulate bone growth and therewith speed up the fracture healing process, an important matter during total hip and knee arthroplasties is prevention of the infections. Recent studies estimate the current incidence of infection to be around 2% to 4% in knee and hip arthroplasties respectively [55-61]. These numbers can increase up to 50%, due to pin track infections, when external fracture fixators are used in the trauma surgery [62,63]. The reason lies in the poor accessibility of the bone-infected site by systematically administered antibiotics. Local therapy is therefore desired and can be achieved by using a suitable carrier. Incorporation of antibiotics into the biomimetically produced coatings is a promising approach for controlled drug delivery and local infection prevention.

In a previously published study by Stigter *et al.* [64], tobramycin ($C_{18}H_{37}N_5O_9$, MW 467.5 g/mol) was incorporated into biomimetic CA coating by adding various concentrations of the antibiotic into SBF-B solution, in which the second step of the coating process is taking place. For comparison, plasma-sprayed HA-coated Ti alloy plates were immersed into similar concentrations of tobramycin for various time periods. The amount of adsorbed antibiotic was measured by using a fluorescent polarizing immunoassay.

The final loadings at same antibiotic concentration in PS-HA coating were 10 times lower than in biomimetic CA coating. Tobramycin could be adsorbed onto the PS HA coating, but the adsorption remained superficial and limited to low amounts. On the contrary, antibiotic molecules were coprecipitated with the crystals of the biomimetic CA coating, throughout the whole layer. Tobramycin did not change the morphology of the coating, but its thickness decreased with the increase of the amount of incorporated antibiotic. The dissolution of the coating and release of tombramycin were measured *in vitro* in SPS at pH5.0 and pH7.3. The release of the antibiotic was gradual, and faster at pH7.3 (50µg/ml/min) than at pH5.0 (4µg/ml/min). Tobramycin released from the biomimetic coating could inhibit growth of *Staphylococcus aureus* bacteria *in vitro*, indicating that the biomimetic CA coating containing antibiotics could be used to prevent post-surgical infections in orthopaedic and trauma.

6. Conclusions

Biomimetic approach for producing CaP coatings on orthopaedic implants offers many possibilities. The physiological conditions of the biomimetic process can broaden the variety of materials to be coated such as heat-sensitive and porous implants, and the variety of CaP phases, including those that are only stable under mild temperature conditions. Biomimetic CaP coatings have shown good biological performance, looking at both their osteointegration and osteoinductive potential. In addition, drugs and growth factors can easily be incorporated into biomimetic coating, further increasing their bioactivity and clinical applicability.

References

1. Wainwright S.A. (1995) What We Can Learn from Soft Biomaterials and Structures in M. and I.A. Aksay (eds.), *Biomimetics. Design and Processing of Materials,* AIP press, Woodbury, US, pp. 1-12.
2. Breme H.J., Biehl V. and Helsen J.A. (1998) Metals and Implants, in J.A. Helsen and H.J. Breme (eds.), *Metals as Biomaterials,* John Wiley & Sons, Chichester, UK, pp. 37-71.
3. Osborn J.F., Newesely H. (1980) Dynamic Aspects of the Implant Bone Interface, in G. Heimke (ed.), *Dental Implants,* Carl Hansen Verlag, Munchen Germany, pp. 111-123.
4. Havelin L.I., Engesaeter L.B., Espehaug B., Furnes O., Lie S.A.and Vollset S.E. (2000) The Norwegian Arthroplasy Register, 11 Years and 73,000 Arthroplasties, *Acta Orthop. Scand.* **71 (4)**, 337-353.
5. Ducheyne P., van Raemdonck W., Heughebaert J.C. and Heughebaert M. (1986) Structural Analysis of Hydroxyapatite Coatings on Titanium, *Biomaterials* **7 (2)**, 97-103.
6. Ong J.L., Lucas L.C., Lacefield W.R. and Rigney E.D. (1992) Solubility and Bond Strength of Thin Calcium Phosphate Coatings Produced by Beam Sputter Deposition, *Biomaterials* **13 (4)**, 249-54.
7. Ben-Nissan B., Chai C.S. and Gross K.A. (1997) Effect of Solution ageing on Sol-Gel Hydroxyapatite Coatings, *Bioceramics* **10**, 175-78.
8. Du C., Klasens P., de Haan R.E., Bezemer J., Cui F.Z., de Groot K. and Layrolle P. (2002) Biomimetic Calcium Phosphate Coatings on PolyActive® 1000/70/30, *J. Biomed. Mater. Res.* **59**, 535-546.
9. Leeuwenburgh S., Layrolle P, Barrère F., Schoonman J., van Blitterswijk C.A. and de Groot K. (2001) Osteoclastic Resorption of Biomimetic Calcium Phosphate Coatings *in vitro, J. Biomed. Mater. Res.* **56**, 208-215.
10. Layrolle P., van der Valk C., Dalmeijer R., van Blitterswijk C.A. and K. de Groot (2001) Biomimetic Calcium Phosphate Coatings and Their Biological Performances, *Bioceramics* **13**, 391-394.
11. Wen H.B., Wolke J.G.C., de Wijn J.R., Cui F.Z. and de Groot K. (1997) Fast Precipitation of Calcium Phosphate Layers on Titanium Induced by Simple Chemical Treatment, *Biomaterials* **18**, 1471-1478.
12. Kim H.M., Miyaji F., Kokubo T. and Nakamura T. (1996) Preparation of Bioactive Ti and its Alloys via Simple Chemical Surface Treatment, *J. Biomed. Mater. Res.* **32**, 409-417.
13. Yamada S., Nakamura T., Kokubo T., Oka M and Yamamura T. (1994) Osteoclastic Resorption of Apatite Formed on Apatite- and Wollastonite-Containing Glass-Ceramic by a Simulated Body Fluid, *J. Biomed. Mater. Res.* **28**, 1357-1363.
14. Li P., Kangasniemi I., de Groot K. and Kokubo T. (1994) Bonelike Hydroxyapatite Induction by a Gel-Derived Titania on a Titanium Substrate, *J. Am. Ceram. Soc.* **77**, 1307-1312.
15. Peltola T., Patsi M., Rahiala H., Kangasniemi I. and Yli-Urpo A. (1998) Calcium Phosphate Induction by Sol-Gel-Derived Titania Coatings on Titanium Substrates *in vitro, J. Biomed. Mater. Res.* **41**, 504-510.
16. Li P. and Ducheyne P. (1998) Quasi-Biological Apatite Film Induced by Titanium in a Simulated Body Fluid, *J. Biomed. Mater. Res.* **41**, 341-348.
17. Habibovic P., Barrère F., van Blitterswijk C.A., de Groot K. and Layrolle P. (2002) Biomimetic Hydroxyapatite Coating on Metal Implants, *J. Am. Ceram. Soc.* **85(3)**, 517-522.
18. ASTM E 376 (1996): "Standard Practice for Measuring Coating Thickness by Magnetic-Field or EDDY-Current (Electromagnetic) Test Methods".
19. ASTM STP 1196 (1994) "Characterization and Performance of Calcium Phosphate Coatings for Implants".
20. NF S94-066 (French standard 1998): "Determination quantitative du rapport Ca/P de phosphates de calcium".
21. Barrère F., Layrolle P., van Blitterswijk C.A. and de Groot K. (1999) Physical and Chemical Characteristics of Plasma-Sprayed and Biomimetic Apatite Coating, *Bioceramics* **12**, 125-128.
22. Barrère F., Layrolle P., van Blitterswijk C.A. and de Groot K. (1999) Biomimetic Calcium Phosphate Coatings on Ti6Al4V: a Crystal Growth Study of Octacalcium Phosphate and Inhibition by Mg^{2+} and HCO_3^-, *Bone* **25**, 107S-111S.
23. Moreno E.C. and Varughese K. (1981) Crystal growth of calcium apatite from dilute solutions. *J. Crystal Growth* **53**, 20-30.
24. Elliot J.C. (1994) *Structure and Chemistry of the Apatites and Other Calcium Orthophosphates,* Elsevier, Amsterdam, The Netherlands.

25. Barrère F., Layrolle P., van Blitterswijk C.A. and de Groot K. (2002) Influence of Ionic Strength and Carbonate on the Ca-P Coating Formation from SBFx5 Solution, *Biomaterials* **23(9)**, 1921-1930.
26. Barrère F., Layrolle P., van Blitterswijk C.A. and de Groot K. (2000) Fast Formation of Biomimetic Ca-P Coatings on Ti6Al4V, *Mat. Res. Soc. Symp. Proc.* **599**, 135-140.
27. Newesely H. (1961) Changes in Crystal Types of Low Solubility Calcium Phosphates in Presence of Accompanying Ions, *Arch. Oral Biol.*, **Special Supplement 6**, 174-180.
28. Tomazic B., Tomson M. and Nancollas G.H. (1975) Growth of Calcium Phosphates on Hydroxyapatite Crystals: The Effect of Magnesium, *Arch. Oral Biol.* **20**, 803-808.
29. Salimi M.H., Heughbaert J.C. and Nancollas G.H. (1985) Crystal Growth of Calcium Phosphates in the Presence of Magnesium Ions, *Langmuir* **1**, 119-122.
30. Eanes E.D. and Rattner S.L. (1980) The effect of Magnesium on Apatite Formation in Seeded Supersaturated Solutions at pH=7.4, *J. Dent. Res.* **60 (9)**, 1719-1723.
31. Boskey A.L. and Posner A.S. (1974) Magnesium Stabilization of Amorphous Calcium Phosphate: A Kinetic Study, *Mat. Res. Bull.* **9**, 907-916.
32. Nancollas G.H., Tomazic B. and Tomson M. (1976) The Precipitation of Calcium Phosphate in the Presence of Magnesium, *Croatia Chemica Acta* **48**, 431-438.
33. Chikerur N.S., Tung M.S. and Brown W.E. (1980) A Mechanism for Incorporation of Carbonate into Apatite, *Calcif. Tissue. Int.* **32**, 55-62.
34. Bachara B.N.and Fisher H.R.A. (1969) The Effect of Some Inhibitors on the Nucleation and Crystal Growth of Apatite, *Calcif. Tissue Res.* **3**, 348-357.
35. Füredi-Milhofer H., Lj. Brecevic and B. Purgaric (1976) Crystal Growth and Phase Transformation in the Precipitation of Calcium Phosphates, *Faraday Discuss. Chem. Soc.* **61**, 184-90.
36. Koutsoukos P., Amjad Z., Tomson M.B. and Nancollas G.H. (1980) Crystallization of Calcium Phosphates: A Constant Composition Study, *J. Am. Chem. Soc.* **102(5)**, 1553-1557.
37. van Kemenade M.J.J.M. and de Bruyn P.L. (1987) A Kinetic Study of Precipitation from Supersaturated Calcium Phosphate Solutions, *J. Colloid and Interface Sci.* **118**, 564-585.
38. Kohman G.T. (1963) Precipitation of Crystals from Solution in Gilman (ed.) *The Art and Science of Growing Crystals*, John Wiley and Sons, London, UK, pp. 152-162.
39. LeGeros R.Z., Kijkowska R. and LeGeros J.P. (1984) Formation and Transformation of Octacalcium Phosphate, OCP a Preliminary Report, *Scan. Electron Microsc.* **IV**, 1771-1777.
40. Cheng P.T. and Pritzker K.P.H. (1983) Solution Ca/P Ratio Affects Calcium Phosphate Crystal Phases, *Calcif. Tissues Int.* **35(4-5)**, 596-601.
41. Iijima M., Kamemizu H., Wakamatsu N., Goto T., Doi Y. and Moriwaki Y. (1994) Effects of CO_3^{2-} Ion on the Formation of Octacalcium Phosphate at pH=7.4 and 37°C, *J. Crystal Growth* **135**, 229-234.
42. Heughebaert J.C. and Nancollas G.H. (1984) Mineralization Kinetics: the Role of Octacalcium Phosphate in th ePrecipitation of Calcium Phosphates, *Colloids and Surfaces* **9**, 89-93.
43. Le Geros R.Z., Daculsi G., Orly I., Abergas T. and Torres W., (1989) Solution-Mediated Transfroamtion of Octacalcium Phosphate (OCP) to Apatite, *Scanning Microsc.* **3(1)**, 129-137.
44. Barrère F., van der Valk C.M., Dalmeijer R.A.J., Meijer G., van Blitterswijk C.A., de Groot K. and Layrolle P. (2003) Osteogenicity of Octacalcium Phosphate Coatings Applied on Porous Metallic Implants, *J. Biomed. Mater. Res.* **66A(4)**, 779-788.
45. Doi Y., Iwanaga H., Shibutani T., Moriwaki Y., Iwayama Y. (1999) Osteocalstic Responses to Various Calcium Phosphates in Cell Cultures., *J Biomed Mater Res.* **47**: 424-433.
46. Yamada S, Heymann D, Bouler JM, Daculsi G. (1997) Osteoclastic Resorption of Calcium Phosphate Ceramics with Different Hydroxyapatite/beta-Tricalcium Phosphate Ratio, *Biomaterials* **18**, 1037-1041.
47. Habibovic P., Li J.P., van der Valk C.M., Meijer G., Layrolle P., van Blitterswijk C.A. and de Groot K. (2003) Biological Performance of Novel Porous Ti6Al4V, *submitted for publication*.
48. Yuan H., de Bruijn J.D., Dalmeijer R.A.J., Layrolle P., van Blitterswijk C.A., Zhang X. And de Groot K. (2001) Bone Induction through Physicochemistry in H. Yuan *Osteoinduction of Calcium Phosphates*, CIP-Data Koninklijke Bibliotheek, The Hague, The Netherlands, pp.123-131.
49. Du C., Meijer G.J., van der Valk C.M., Haan R.E., Bezemer J.M., Hesseling S.C., Cui F.Z., de Groot K. and Layrolle P. (2002) Bone Growth in Biomimetic Apatie Coated Porous PolyActive® 1000PEGT70PBT30 Implants, *Biomaterials* **23**, 4649-4565.
50. Habibovic P., van der Valk C.M., Meijer G., Layrolle P., van Blitterswijk C.A. and de Groot K. (2003) Influence of Octacalcium Phosphate Coating on Osteoinductive Properties of Biomaterials, *submitted for publication*.

51. Liu Y., Hunziker E.B., Randall N.X., de Groot K. and Layrolle P. (2003) Proteins Incorporated into Biomimetically Prepared Calcium Phosphate Coatings Modulate their Mechanical Strength and Dissolution Rate, *Biomaterials* **24**, 65-70.

52. Wen H.B., de Wijn J.R., van Blitterswijk C.A. and de Groot K. (1999), Incorporation of Bovine Serum Albumin in Calcium Phosphate Coating on Titanium, *J. Biomed. Mater. Res.* **46(2)**, 245-252.

53. Liu Y., Hunziker E.B., Layrolle P., and de Groot K. (2002) Introduction of Ectopic Bone Formation by BMP-2 Incorporated into Calcium Phosphate Coatings of Titanium-Alloy Implants, *Bioceramics* **15**, 667-670.

54. Schenk R.K., Hunziker E.B. (1994) in C.T. Brighton, G.E. Friedlaender and J.M. Lane (eds.) *Bone Formation and Repair*, The American Academy of Orthopaedic Surgeons, Rosemont, IL, 325-340.

55. Fitzgerald R.H. (1992) Total Hip Arthroplasty Sepsis, *Orthop. Clin. North. Am.* **23(2)**, 259-264.

56. Nasser S. (1992) Prevention and Treatment of Sepsis in Total Hip Replacement Surgery, *Orthop. Clin. North. Am.* **23(2)**, 265-277.

57. Norde C.W. (1991) Antibiotic Propylaxis in Orthopaedic Surgery, *Rev. Infect. Dis.* **13(10)**, 265-277.

58. Espehaug B.and Egesaeter L.B. (1997) Antibiotic Propylaxix in Total Hip Arthroplasty, *J. Bone Joint. Surg.* **79-B**, 590-595.

59. Schmalzried T.P., Amstutz H.C., Au M.K. and Dorey F.J. (1992) Etiology of Deep Sepsis in Total Hip Arthroplasty, *Clin. Orthop. and Rel. Res.* **280**, 200-207.

60. Schutzer S.F. and Harris W.H. (1988) Deep-Wound Infection after Total Hip Replacement under Contemporary Aseptic Conditions, *J. Bone. Joint Surg.* **70**, 724-727.

61. Wilson M.G., Kelley K. and Thornhill T.S. (1990) Infection as a Complication of Total Knee Replacement Arthroplasty, *J. Bone. Joint Surg.* **72(6)**, 878-883.

62. Green S.A. and Ripley M.S. (1984) Chronic Ostemyelitis in Pin Tracks, *J. Bone Jt. Surg.* **66A**, 1092-1098.

63. Green S.A. (1983) Complications of External Skeletal Foxation, *Clin. Orthop. Rel. Res.* **180**, 109-116.

64. Stigter M., de Groot K. and Layrolle P. (2002) Incorporation of Tobramycin into Biomimetic Hydroxyapatite Coating on Titanium, *Biomaterials* **23**, 4143-4153.

LEARNING FROM NATURE HOW TO DESIGN BIOMIMETIC CALCIUM-PHOSPHATE COATINGS

I. B. LEONOR[1,2], H. S. AZEVEDO[1,2], I. PASHKULEVA[1,2], A. L. OLIVEIRA[1,2], C. M. ALVES[1,2], R. L. REIS[1,2]

[1]*3B's Research Group - Biomaterials, Biodegradables and Biomimetics, University of Minho, Campus de Gualtar, 4710-057 Braga, Portugal*
[2]*Department of Polymer Engineering, University of Minho, Campus de Azurém, 4800-058 Guimarães, Portugal*

1. Introduction

1.1. NATURE INSIGHTS ON BIOMINERALIZATION

Mineralized biosystems can gather unique extraordinary properties that most of the times have a complexity beyond our knowledge. However, their formation follows general common principles across different species. In contrast to colloidal or solution processing techniques commonly used for the production of synthetic minerals or their precursors, biomineral production occurs under moderate conditions of supersaturation [1]. For biomineralization to occur specific subunit compartments or microenvironments need to be created, in order to stimulate crystal formation at certain "functional sites" and inhibition or prevention of the process at all other sites [1-3]. The highly specific control of morphology, location, orientation and crystallographic phase all indicate the existence of an optimized or "engineered" substrate surface. The key characteristics of these optimized interfaces are a mystery at present, namely because of the complexity of most biological model systems. Investigations of representative systems, such as nacre [4,5], dentin [6,7], enamel [8-10], cartilage [11,12] and bone [13-15], highlight the definite importance of the microenvironments and of the orientation between the organic matrix and the precursors for the mineral formation. At the macroscopic level the growth of a biomineralized structure gains shape by the sequential packaging of units together, resulting in unique composite structures that are prepared to accommodate later stages of the organism growth and repair [16]. The sophistication of nature's "bottom-up" approach for the production of complex but yet functional structures has inspired a great number of researchers in order to fabricate enhanced materials. However, to achieve Nature's efficacy on solving its problems is still a utopia. Nevertheless, the richness of information that is stored in each biomineralized tissue has been a source of new ideas over different fields like chemistry

R.L. Reis and S. Weiner (eds.),
Learning from Nature How to Design New Implantable Biomaterials, 123-150.
© 2004 *Kluwer Academic Publishers. Printed in the Netherlands.*

[3,17-20], biochemistry [21-24], materials science and engineering [18,19,25-27], materials design [28,29] and even architecture or art [30] (See for example, Figure 1).

Figure 1. Inspiration from nature: the internal skeletal of a shell and stairs in one of the towers in the cathedral "La Sagrada Família", in Barcelona, Spain.

Progresses are being made on the process of learning from biomineralized tissues in which the formation of hierarchical structures is present. These complex architectures, going from the nanometre to the millimetre scale, combine minerals, structural biological polymers (proteins and polysaccharides), lipids and ultimately cells [13]; under the correct genetic control of the shape and pattern in time and space, the biocomposites are produced with minimum energy costs using an environmentally friendly biomineralization synthesis pathway [25,31]. The resulting tissue tends to optimise its function, many times its multifunctions, preferentially minimizing the amount of materials. Moreover, biological materials are also "smart" as they adapt internally to external events, which includes some extent of "self-repairing" ability.

1.2. EVENTS ON BONE MINERALIZATION

Bone is among the most complex examples of a biomineralized material [13]. Although it has been subject of a lot of research in the past few decades, the fact is that the complete mechanism for bone tissue formation still raises a number of questions which still await satisfactory answers.

In the early days, mineralization was visualized as being a physicochemical phenomenon in which mineral nucleation is achieved by extracellular, non-living chemical structure(s) in the matrix (e.g., collagen) that would act as templates upon which the first mineral crystals were formed [32]. Initial crystals could then serve as nuclei for further mineral propagation. In this perspective, both initiation and propagation were controlled by non-living chemical factors residing in the matrix. Physicochemical studies have then determined the solution conditions that favour the direct precipitation of different calcium phosphate and other calcium containing phases. These studies could easily explain different phenomena of pathologic calcification such as: why in the acidic environment of the mouth, brushite may appear in dental calculus

[33], or why pancreatic stones formed in a carbonate-rich environment are calcium carbonates [34]. However, an old question that has created a great deal of controversy ever since is why the mineral of bone forms at specific sites at the matrix. What factors control the process? These questions were raised some decades ago and are still contemporary.

Over the last 30 years, the common opinion started to shift to a view that envisioned cells as being importantly involved in mineral initiation while mineral propagation would remain primarily extracellular and physicochemical (collagen-mediated) [14,15,35-39], but always regulated by cells through the creation of a matrix and an ionic milieu in which mineralization may or may not progress. The mechanism of cell-mediated mineralization started to be best visualized as a cascade of events requiring the interaction of many different factors that either promote or retard/inhibit this phenomenon. The whole mechanism and the specific role of each factor in particular have not yet been completely elucidated. However, a widely accepted mechanism seems to be based on the formation of matrix vesicles (MV), extracellular membrane-invested and lipid-rich bodies released by budding from the surfaces of osteoblasts [14,15,35-38]. Active transport (ion pumps) may be used to raise the amount of calcium and phosphate in the vesicles to levels above supersaturation, creating favourable ionic conditions for deposition of nascent mineral within the protected microenvironment of the MV membrane. Initially the mineral is in the form of amorphous calcium phosphate, octacalcium phosphate and/or brushite, with later conversion to hydroxyapatite [36]. The membrane of the vesicles provides a protected microenvironment in which Ca^{2+} and PO_4^{3-} can be concentrated, localized and interact to form the first, unstable, nuclei of mineral. Alkaline phosphatase activity (and the activity of other vesicle phosphatases) functions to promote mineralization during this phase, as do the Ca-binding phospholipids and proteins of the MV [37]. Annexins could then serve as Ca^{2+} ion channels (as they secure free calcium), promoting Ca^{2+} transport to the initial site of mineralization, inside the MV membrane [38,40,41]. Once these nuclei are transformed into HA, mineral propagation begins with the penetration of the MV membrane by crystals, and their exposure to the extracellular fluid (ECF) (phase 2) [14,15,35-38]. Crystal exposure is promoted by proteinases and lipases in MVs that speed vesicle breakdown [38]. In the absence of apatite, ECF contains insufficient Ca^{2+} and PO_4^{3-} to initiate mineral deposition, but a sufficient amount is present to support crystal proliferation. The preformed apatite crystals from the MV act as templates for new crystal proliferation. Molecules at the outer MV surface such as collagen type X may serve as a bridge for mineral to spread into the adjacent collagenous matrix [38].

A good example of this biological strategy of using lipid vesicle compartments to sequester ions and control mineralization is found, in a much more simple biomineralized organism, in the cytoplasm of magnetoestatic bacteria [42], were a string of nanosized magnetic iron oxide particles is deposited within pre-existing lipid vesicles. The individual magnetite particles (typically 100 nm diameter) are organized with their magnetic axes aligned in the same direction, and the entire construct is believed to be used by the organism to navigate according to the magnetic field of the earth. Therefore, phospholipids are important contributors to mineralized tissue formation. Eanes [43] have used synthetic liposomes as tools to model the mineral precipitation sequence level to occur during matrix vesicle mineralization. Phosphate-loaded liposomes of size similar to MVs were prepared by encapsulation of aqueous inorganic phosphate solutions within liposomes constructed from phosphatidylcholine, dicetyl phosphate and cholesterol. The lipid bilayer which serves as a barrier to ion transport, isolated extraliposomal Ca^{2+} from entrapped phosphate and prevented mineral formation. However, when these liposomes were made permeable to external Ca^{2+} using calcium inophore, apatite mineral readily formed inside and then growth through the external medium, penetrating the lipid bilayer membrane. This model system was also used to investigate the role of acidic phospholipids.

The diversity of organic and inorganic substances that can actually affect all stages of bone development is enormous and yet to be fully understood. Not only do cells directly influence the size, rate, and type of mineralization thought the synthesis and strategic placement of MVs, but they also regulate the mineralization indirectly thought the production of a variety of molecules. In fact, cell-mediated mineralization requires the interaction of a number of other factors that either promote or retard/inhibit this phenomenon. As recently presented in a review by Sikavitsas et al. [44]some of the most important molecules can be divided generally into soluble factors and attached matrix molecules and their effects are briefly presented here, in Table 1 [14,15,35,37,38,44-65].

Several inorganic substances can also influence bone formation. Metal ions like iron (III) indium and strontium have shown to have a stimulatory effect on the conversion of amorphous Ca-P to HA [66]. On the contrary, magnesium, aluminium and in this order, Ni^{2+}, Sn^{2+}, Co^{2+}, Mn^{2+}, Cu^{2+}, Zn^{2+}, Ga^{3+}, Tl^{3+}, Mo^{5+}, Sb^{3+}, have shown an inhibitory effect on apatite formation [66].

1.2.1. Templates for mineralization

Besides the biologically based events and the role of the organic/inorganic substances that contribute and control the mineralization phenomena, other aspects are equally important in this complex process. The highly specific control of morphology, location, orientation and crystallographic phase of the mineral all indicate the existence of an optimized or "engineered" substrate surface. In fact, mineral crystals deposit on collagen fibrils and initial crystals deposit at discrete sites on these fibers. This means that collagen has a role in controlling mineralization. But is there more than collagen (other non-collageneous proteins), is it a specific type of collagen? Therefore, what is the relevance of an organic template? How the specificity of the surface really contributes for the mineralization?

In Nature, there are many examples of structures where minerals are complexed with organic molecules to form hybrid materials, including the structure of vertebrates laminar bone [19,26,28,67,68], dentin [26,28,68,69], mollusk shells [26,28,69,70], composite fibrils formed by biogenic silica deposition in plants [28,67], the membrane of some bacteria [67], etc.

TABLE 1. Examples of molecules influencing stages of bone development [14,15,35,37,38,44-65]

Matrix components	Brief description of the effect on the process of bone mineralization	Relevant References
Alkaline phosphatase	A cell-linked polypeptide, an enzyme, that is secreted by osteoblasts. It is localized in the MVs and is thought to promote crystal formation by removing nucleation inhibitors. It is inhibited by glucocorticoids and parathyroid hormone	[14,35,37,38]
Osteocalcin	A noncollagenous protein which as a moderate affinity for calcium and may be used to inhibit mineralization. It could also play a role in bone resorption.	[57,62]
Osteonectin	A glycolprotein that binds to calcium, hydroxylapatite, and collagen, suggesting that it is a nucleator for the matrix mineralization. It also binds to trombospondin, and its synthesis is stimulated by vitamin D.	[61-63]
Thrombospondin	A trimeric glycoprotein that is secreted by connective tissue cells. It binds calcium, hydroxylapatite, and osteonectin. It can also attach to bone and cells. It may organize extracellular matrix components or act as growth factor.	[64]
Fibronectin	An extracellular polypeptide with binding regions for collagen, fibrin, as well as cells. Osteoblasts use fibronectin for cell attachment.	[38,53-55]
Proteoglycans I and II	Both may affect collagen growth and diameter of the fiber.	[15,38,49]
Osteopontin	An acidic sialoprotein (protein conjugated with sialic acid), it is implicated in general cell attachment to the bone matrix.	[38,46,48]
Bone sialoprotein	Another sialoprotein, it also has cell attachment activity but maintains cell attachment for shorter periods than osteopontin.	[46-48]
SOLUBLE FACTOR		
Bone morphogenic proteins (BMP)	A family of cytokines that stimulates proliferation of both chondrocytes and osteoblasts and causes increased matrix production in each cell type. BMPs have also been seen to induce MSC differentiation to osteoblasts.	[35,50,51]
Platelet-derived growth factor (PDGF)	A growth factor, it stimulates proliferation of chondrocytes and osteoblasts. However, in different concentrations, it has also been implicated in bone resorption.	[65]
Transforming growth factor-β (TGF-β)	A growth factor, it causes differentiation of MSCs to chondrocytes, and may also induce chondrocyte and osteoblasts proliferation. Since, like PDGF, it has been seen to enhance bone resorption at certain concentrations, it may play a role in coupling formation and resorption activities.	[38]
Fibroblast factors (FGF)	A family of growth factors that stimulates proliferation of MSCs, osteoblasts and chondrocytes.	[57]
Insulin-like growth factors (IGF)	A family of growths that stimulates proliferation of chondrocytes and osteoblasts as well as induces matrix secretion from both cell types.	[52]
Parathyroid hormone (PTH)	A single chain polypeptide, it causes the release of calcium from the bone matrix as well as induces osteoclast differentiation from precursor cells. It is also thought to inhibit osteoblasts function.	[35,58]
Estrogen	A hormone, it has a complex effect on bone, with the final outcome being decreased bone resorption by osteoclasts.	[59]
Calcitonin	A polypeptide, it lowers levels of blood calcium by inhibiting osteoclasts function.	[58,60]
Prostaglandins	A family of fatty acids that, while initially inhibiting osteoclasts, over extended periods encourages the proliferation and differentiation of these cells.	[35,71]
Interleukin-1 (IL-1)	Stimulates the proliferation of osteoclasts precursors, thus increasing bone resorption.	[72]
Vitamin D	Has a complex effect on bone formation, possibly by regulating the synthesis of other molecules. It has been reported to influence both bone resorption and matrix mineralization.	[45,60]

Weiner and colleagues [28] believe that the way by which the mineral phase and the organic material are organised is the key factor in contributing to the distinct mechanical properties of these biological materials. Having into account these natural phenomena, many researchers [67,73-75] have been using different protein molecules to control the nucleation and growth of mineral phases and consequently manipulate their properties (composition, structure, crystallinity, morphology, mechanical strength).

The major role of these molecules may be either templating or enzymatic effects [69]. A macromolecular template could provide stereochemistry and physical adsorption for the inorganic formation. On the other hand, an enzyme could regulate inorganic phase synthesis by controlling local chemistry [69]. The entrapment of biomacromolecules in mineral phases may also contribute to strength materials [67,76] and also for stabilizing the mineral contents [76,77]. For instances, Liu and co-workers [76,78,79] reported that the dissolution of calcium phosphate (Ca-P) coatings on the surface of titanium implants was reduced when bovine serum albumin was incorporated into the coating. It was postulated that the protein was able to reduce the release of Ca from the coating and to enhance the adhesion between the coating and titanium [79].

A good case study for finding interesting clues occurs in shells of bivalves that exhibit two layers of differently shaped aragonite crystals, neither of them being similar to the crystals formed inorganically [80]. This is due to the precise calcium-binding sites that are dictated by the structure of the nucleating proteins (negatively charged, interacting electrostatically with the calcium ions), which are glycoproteins rich in aspartic acid, possibly with a β-sheet structure [29,80]. Such stereochemical self-assembled templates will control both the nucleation and the growth of the inorganic precipitation reactions, determining the final size, shape and orientation of the formed crystalline structure. These "control macromolecules" are usually the minor macromolecular component of a biological material, being intimately attached to the mineral phase; they are, in fact, difficult to extract or degrade without dissolving the mineral. It is interesting to notice in this context that such acidic soluble proteins inhibit crystallization in solution, as rigidity and regularity is then lost. This is related with the way how Nature controls the biomineralization process: minerals formed in organisms require a precise isolation in space, either delineated by macromolecular matrix frameworks (providing a 3D matrix for crystal formation and a substrate for the interaction between the control macromolecules and the mineral phase [29]), or either by confinement from cell membranes or vesicles [3,38,80]. These evidences suggest that analogous synthetic templates could be used in the production of tailor-made mineral structures. As referred by Green et al. [81], surfactant micelles, lipid microstructures, and stacked bacterial filaments have been proposed to control precipitation and growth of inorganic mineral's such as silica and iron oxides. In that particular work, an analysis was made where chemical and biological concepts of self-assembling and self-organization found in natural porous skeletons can be used in the development of bone-analogue structures for hard tissue engineering [81]. It should also be noticed in this context that apparently different biomineralized structures can exhibit physiological compatibility [82].

The role of proteins carrying sequences of carboxylates in the process of crystal growth of biominerals such calcium phosphates, oxalates and carbonates has also been extensively investigated [83-85]. The aspartic (Asp) and glutamic (Glu) amino acids appear to play an important role in the process of protein binding on the crystal surface.

For example, osteonectin appears to bind on hydroxyapatite surfaces through a glutamate-rich sequence and bone sialoprotein (BSP) has regions rich in Glu aminoacids which are also involved in bone mineralization [84]. The importance of protein carboxylic (-COOH) side chains in mineralization has been tested by researchers using model systems [83,85]. Carboxylate-containing molecules have been studied to induce the nucleation and growth of various minerals. The works of Addadi *et al.* [83,85]have shown that polymerized and self-assembled monolayers of carboxylates have been successfully used for nucleation of calcite.

Osteopontin, bone sialoprotein, and bone acidic glycoprotein-75 are three acidic phosphoproteins that are isolated from the mineralized phase of bone matrix, are synthesized by osteoblastic cells, and are generally restricted in their distribution to calcified tissues [46]. Although each is a distinct gene product, these proteins share aspartic/glutamic acid contents of 30-36% and each contains multiple phosphoryl and sialyl groups. These properties, plus a strict relationship of acidic macromolecules with cell-controlled mineralization throughout Nature, suggest functions in calcium binding and nucleation of calcium hydroxyapatite crystal formation. However, direct proof for such roles is still largely indirect in Nature. A two-state conformational model of calcium binding by bone matrix acidic phosphoproteins was proposed by Gorski [46] based on speculative hypotheses regarding acidic phosphoprotein function.

1.2.2. *The potential for mimicking bone*

Research on biological and synthetic models has been, no doubt, the best approach for finding the ideal surface for mineral formation. When bringing this knowledge to the field of bone replacement materials, better solutions for advanced materials are expected to be found. For instances, the study of the surface functional group dependence of apatite mineral to be formed on self-assembled monolayers *in vitro* has given interesting clues on the physicochemical processes associated with bone formation. This subject will be discussed in more detail in Section 3 of the present chapter, since it is evident that although progresses are being made, acellular studies are still contributing with useful information for better understanding of mineralization and there is still a lot of work to be done in this area. On the other hand, it is clear that the concept of biomimetics also envisages the use of inorganic surface specific proteins as templates or enzymatic agents for controlling materials assembly [69]. The traditional approach involves protein isolation and purification, amino acid analysis and sequencing, which may be complex and time-consuming procedures. Therefore, an alternative approach is to use existing proteins that may be able to modulate the structure of inorganic surfaces, but also for regulating other surface interactions, such as cell attachment, growth and differentiation, by presenting bioactive proteins to the interface.

The role of several organic molecules or inorganic compounds is determinant in endorsing a positive effect for stimulation of bone production. Therefore, the isolation procedures of some of these essential substances have been investigated, optimised and "adopted" by the pharmacologic field as they are effective on the treatment against bone dieses like osteoporosis, osteoarthritis, Paget's disease, etc [58]. However the pharmacokinetics and the side-effects of the administration method (usually oral) of some of these drugs are far from ideal. Nowadays, other applications are envisioned for these "bioactive" factors. They are useful for Tissue Engineering applications [86] and/or to be loaded in carriers for drug release [87]. With the advances in these research areas a new generation of materials is emerging, with tailored properties (sometimes at the nano level) able to respond to the specific stimulus of the local environment. The outermost surface of a material plays, therefore, a crucial role after implantation, defining the type of interface with bone. Since bone metabolism at the surface of implants is not always satisfactory, there is still a need for improvements regarding the stimulation of bone formation at the implantation site. This may be achieved by delivering osteoinductive factors (being these organic or inorganic) at the interface, able to induce local bone growth around the implant. The idea of the following sections of this chapter is, therefore, to propose different choices for a biomimetic design of the interface biomaterial/bone, particularly when using previously proposed natural origin biodegradable materials [88-92]. Possible approaches presented herein go from coating the surfaces with a bioactive bone-like apatite layer through a biomimetic methodology to the modification of the surfaces with certain functional groups that are known to induce apatite formation.

2. Biomimetic Calcium Phosphate Coatings

Design strategies for creating Ca-P coatings that have a dual beneficial effect, by one side its osteoconductive properties and on the other side a carrier potential, might have a very promising future. The 3B's Research Group at the University of Minho, has been studying the incorporation of proteins and enzymes onto Ca-P coatings, generated by a biomimetic route [93,94], with the aim of producing coatings with novel properties (in terms of morphology, crystallinity, stability, mechanical strength) and also as a delivery system of therapeutic agents.

It is known that the formation of an apatite layer on the surface of biomaterials is an essential requirement for an implant to exhibit a bone-bonding behaviour [95]. Therefore, many coating techniques have been developed to produce biocompatible and osteoconductive surfaces able of guiding bone formation. The use of Nature inspired methods, such as biomimetic coatings, presents the advantage of using physiological conditions, like mild temperature and pH. This, in turn, allows the incorporation of bioactive species without compromising their performance and to improve the functionality of the inorganic layer at the implant interface.

The concept of biomimetics has been explored by several authors [26,30,69,95-99] under different perspectives. Based on this concept, Kokubo et al. [97] developed a technique for coating different organic, inorganic and metallic materials, with bioactive layers, which has been designated as biomimetic coating. The main aim of this biomimetic process is to mimic the biomineralization, leading to the formation of a bone-like carbonated apatite layer on the surface of a substrate. The methodology has been claimed to be very useful for producing highly bioactive and biocompatible composites with different mechanical properties [100-103].

The original biomimetic coating methodology includes two steps which can be summarized as follows [97,104]: the substrates are placed typically near a $CaO-SiO_2$-based glass (MgO 4.6, CaO 44.7, SiO_2, 34.0 P_2O_5 16.2, CaF_2 0.5 wt%) particles immersed in a simulated body fluid (SBF) [105] solution with ion concentrations (Na^+ 142.0, K^+ 5.0, Mg^{2+} 1.5, Ca^{2+} 2.5, Cl^- 147.8, HCO_3^- 4.2, HPO_4^{2-} 1.0, SO_4^{2-} 0.5 mM) nearly equal to those of human plasma at 36.5 °C. The glass particles release large amounts of calcium and silicate ions, which are adsorbed onto the surface of the substrate to induce the apatite nucleation, and the calcium ions increase the degree of supersaturation with respect to apatite in the SBF, which accelerates apatite nucleation (this first period is designated as nucleation stage). To allow the growth of the apatite nuclei formed on the substrate in the first stage and the formation of an apatite layer, the substrate is immersed in another solution e.g. 1.5 SBF (Na^+ 213.0, K^+ 7.5, Mg^{2+} 2.3, Ca^{2+} 3.8, Cl^- 223.2, HCO_3^- 6.3, HPO_4^{2-} 1.5, SO_4^{2-} 0.8 mM) with ion concentrations 1.5 times those of the SBF at 36.5 °C (second period designated as growth stage). The thickness of this apatite layer increases as function of the immersion time, and the growth rate of the apatite layer increases with the increment of the ion concentrations 1.5 SBF solution [97,105]. More details on the evolution of several Ca-P coating methodologies over the last years can be found elsewhere [106].

Although very popular and effective, the "traditional" biomimetic process, using bioactive particles as nucleating agents, still present some difficulties on what concerns to the adhesion of the apatite layer to polymeric surfaces as well as on coating materials with complex shapes [107]. Therefore, an adaptation of this biomimetic methodology was made by Reis et al. [96], in which the samples were rolled on a bed of wet bioactive glass particles before immersion in an SBF solution. The methodology was successful on coating different types of polymers and shapes like a high molecular polyethylene, a biodegradable starch poly (ethylene vinyl alcohol) blend (SEVA-C) and a polyurethane foam. However, problems associated with some lack of coating adhesion were still also observed (although better results than for the original method could be obtained). To overcome this problem, different surface treatments were experimented by Oliveira et al. [108], prior to immersion in SBF, like potassium hydroxide (KOH), acetic anhydride, UV radiation and overexposure to ethylene oxide sterilization (EtO), on SEVA-C substrates. New biomimetic methodologies were then developed by the 3B's Research Group based on different approaches like: "impregnation" with a sodium silicate gel [109,110], pre-coating with a calcium silicate layer [111] or incubation in several supersaturated salt solutions ($CaCl_2$, KCl and $MgCl_2$) [112]. These surface treatments were performed prior to immersion in a simulated body fluid (SBF), in order to generate nucleating sites for the formation of the apatite layers. The developed methodologies were aimed at: (i) the reduction of the incubation periods for apatite formation; (ii) the improvement of the adhesion strength between the coating and substrate; (iii) the production of Ca-P layers with different (tailored) Ca-P ratios. (iv) coating the inside of pores in porous 3D architectures to be used on tissue replacement and as tissue engineering scaffolds. Particularly silicate based methodologies have demonstrated to be extremely effective on coating porous scaffolds with different morphologies. Since gels were used, it was possible to cover inside the cell walls of different porous structures [109,111]. Recently, another innovative coating methodology to produce an apatite layer is being proposed by Leonor et al [113]., based on an auto-catalytic deposition route. This original approach uses a deposition route that is totally "electroless", i.e., does not require the use of electric current for its application, being based on redox reactions. Two types of solutions are being studied, alkaline and acid baths, to produce the novel auto-catalytic Ca-P coatings. The developed route seems to be a very promising and simple methodology for being used as a pre-implantation treatment to coat several types of materials previously to their clinical application.

2.1. BIOMIMETIC Ca-P COATINGS AS CARRIERS FOR BIOACTIVE PROTEINS

The functionalization of biomaterial surfaces with bioactive proteins to induce specific cell and tissue responses (e.g. proliferation and/or differentiation) has received great attention in the last years. In fact, the immobilization of biologically active molecules on the surface of biomaterials for presenting effectors to target cells, or to induce a particular effect, is of great interest, since the immobilization of active molecules presents the advantage of providing a continuous and localized stimulus for cells. Unlike non-immobilized active agents, which are often consumed by cells, immobilized biomolecules remain bound to a substrate that is not consumed and thus remain available to stimulate additional cells [114]. Furthermore, in biomedical applications, the efficacy of immobilized biomolecules for stimulating specific cell responses, depends on the mode by which these modulators are presented to the target cells. In this case, it is important to control the densities of immobilized proteins, the binding strength and mostly important their binding orientation, being necessary to ensure the full bioactivity and the correct orientation of the molecules after immobilization [114].

There are many methods for protein immobilization and most of them are based on simple monolayer depositions [115] including covalent binding [116,117], physical adsorption [117,118] or molecular self-assembly, based on chemical and affinity adsorption [115,117], among others [117]. All these methods present advantages and drawbacks regarding protein immobilization efficiency and retained activity/functionally of the immobilized protein. Some of these techniques require chemical modification of the matrix, which may result in the material degradation, especially when biodegradable polymers are used. In addition, these modifications, necessary to attach the protein to the matrix, often result in the loss of protein functionality/activity.

Ca-P coatings, produced by biomimetic routes, offer an opportunity to incorporate protein molecules without compromising its functionality/activity, as they are generated under physiological conditions. Bioactive proteins can be directly integrated in the structure of Ca-P coatings without any covalent bonding with the inorganic phase and maintain their secondary structure close to their native form.

The local degradation of the coating would allow the progressive release of multiple associated active agents directly to the interface, endowing biomimetically prepared coatings of value as controlled drug-release systems [79,115,118].

2.2. INCORPORATION OF PROTEINS AND ENZYMES ONTO BIOMIMETIC Ca-P COATINGS: HOW THESE BIOMOLECULES AFFECT THE *IN VITRO* MINERALIZATION PROCESS?

Our research group has been studying the incorporation of different proteins (bovine serum albumin and α-amylase) at different stages of the preparation of biomimetic Ca-P coatings (nucleation or growth stage) and how the addition of these protein molecules affects the properties of the coatings [93,94,119]. Serum albumin has been used as a model protein due to its concentration in the blood plasma, binding and diffusion properties [120-122]. On the other hand, α-amylase has been studied in terms of enzymatic properties and ability to interact with starch-based materials [123].

Two main questions need to be considered in such type of study: (i) how does the incorporation of protein molecules affect the properties of the coatings? (ii) is this methodology suitable to incorporate bioactive molecules without comprising their activity?

The incorporation of proteins in crystals normally leads to changes in the number, size and distribution of crystal aggregates [124] but the extent of these effects depends on some parameters like the protein concentration [73,125,126] and the way these biomacromolecules are presented in solution (free or immobilized) [125,127-129]. As the protein concentration increases, there is a larger surface area and an increased number of functional groups to interact with the growing crystal nuclei, which facilitates crystallite aggregation [73]. At certain concentrations, however, proteins may have an inhibitory effect [126].

Differences in the amino acid composition, and in particular the number and orientation of acidic groups exposed at the protein surface (protein conformation) [125], also affect the mineral formation. These functional groups may be involved in the initial stages of mineral formation through complexation of calcium cations, which in turn, attract phosphate anions. On the other hand, other molecules prevent the generation of crystal nuclei [130]. As stated along this chapter, some proteins are known to act as nucleators of the mineralization process like, for example, bone sialoprotein and dentine phosphoryn, whereas osteocalcin and osteopontin have an inhibitory effect [124,130].

It was also reported that, when present in solution, some proteins were capable of inhibiting the formation of calcium phosphates, either by binding calcium or phosphate [131] or by adsorbing onto apatite surfaces [131], thus blocking active growth sites. The immobilization of those proteins (adsorbed as a film or bound to a support material) showed, on the contrary, to initiate mineral formation [127,132]. It is believed that, when the protein is adsorbed, it will complex with calcium at the surface, inducing precipitation, whereas the complexation in solution retards the precipitation. Campbell and Nancollas [131] investigated the ability of various salivary proteins to mineralise hydroxyapatite, when immobilised as films, and found that salivary amylase was shown to be the most active protein in inducing the nucleation and growth of calcium phosphate.

It has been suggested that there are two general types of Ca-P coatings inhibitors [126]: those affecting nucleation and those affecting growth, since both these mechanisms are quite different. A protein may act in different ways against crystal nucleation and growth. The effect of protein adsorption on the nucleation stage is complex since protein adsorption may occur on the substrate and on the Ca-P nuclei.

134

Figure 2 shows micrographs, obtained by scanning electron microscopy (SEM), of Ca-P coatings grown on the surface of starch-based polymer (50/50 wt.% polymeric blend of corn starch with poly(ethylene-vinyl alcohol) copolymer, designated by SEVA-C) under different conditions. Further details on the material processing and respective mechanical properties can be found elsewhere [89,133]. In these micrographs it is possible to observe Ca-P coatings with distinct morphologies and crystal aggregates with different sizes. The addition of proteins seems to increase the nuclei size, but when incorporated in the growth stage (Figure 2.c) this effect is more evident and the nuclei appear to be denser. It is well known that the formation of nuclei is a critical step for the development of a uniform and dense mineral layer. The adsorption of proteins on the Ca-P nuclei during the nucleation stage may have retarded their growth and consequently their size is smaller (Figure 2.b). In another study developed by Combes *et al.* [122] it was reported that the adsorption of albumin decreases the interfacial energy of the Ca-P nuclei and therefore stabilizing nuclei with smaller radii.

It is also possible to observe that the proteins seem to be part of the coating, forming a composite structure, and are not merely adsorbed.

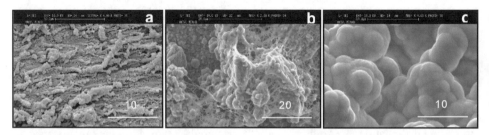

Figure 2. Morphology of Ca-P coating grown on the surface of a starch polymeric blend (SEVA-C) using a biomimetic technology without (a) or with BSA and α-amylase incorporated at concentration of 1mg/mL in the nucleation stage (b) or in the growth stage (c).

Furthermore, an important consequence of growth inhibition in the presence of certain additives is that the crystal morphology can be dramatically changed. Low and high molecular weight additives can induce modifications on the crystal morphology by changing the relative growth rates of different crystal faces through molecular-specific interactions with particular surfaces [3,70]. Electrostatic, stereochemical and structural matching are important factors that significantly modify the surface energy or mechanism of growth, or both. Fast growth along one axis alone gives rise to a needle-shaped crystal, whereas fast growth along two directions produces a plate-like morphology and equal rates of growth in all directions yield isotropic habit (crystal morphology) such as cube or an octahedron [3]. For example, it has been observed that the presence of monosaccharides, like glucose, induced the precipitation of needle-shaped hydroxyapatite crystals elongated along the c-axis [68]. Figure 3 displays the Thin-Film X-ray diffraction (TF-XRD) patterns of Ca-P coating produced on the surface of SEVA-C in the presence of protein molecules (BSA and α-amylase). It can be seen that there is the appearance of well-defined peaks with specific orientation. Such peaks were not, however, observed for the control condition, in the absence of proteins (data not shown). This may indicate some effects of the protein molecules on the orientation of crystal growth. On the other hand, the presence of sugars in solution, which are released from starch hydrolysis catalysed by α-amylase, may also contributed for the crystal growth in a preferential orientation. In fact, it was detected the production of reducing sugars (Figure 4) during the nucleation and growth stages and the presence of these additives may had influenced the crystal growth orientation [68].

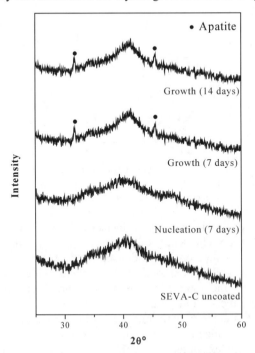

Figure 3. TF-XRD patterns of the polymer surface during the various stages of the Ca-P layer formation using a biomimetic technology with BSA and α-amylase incorporated at concentration 1 mg/mL in the growth stage.

α-amylase is an endo-specific enzyme which catalyses the hydrolysis of α-1,4-glycosidic linkages of starch to maltose and dextrins, reducing the molecular size of starch. The activity of α-amylase can be monitored by measuring the concentration of reducing sugars liberated into the solution. In Figure 5, the concentration of reducing sugars produced after 7 days of incubation in the nucleation and growth stages are compared. It can be observed that, for the same incubation time and enzyme concentration, higher concentration of reducing sugars was detected when the enzyme was incorporated in the growth stage. Nucleation and growth stages are always quite distinct events in the mineralization process. Nucleation is characterised by a series of reactions occurring between the glass particles (Bioglass®) and the SBF solution, leading to the formation of nuclei, and this may had affected the enzyme activity. On the other hand, enzyme molecules might have adsorbed on the glass particles and this had led to some enzyme denaturation. This result indicates that the incorporation of proteins at the growth stage is preferred, especially when it is desired to retain the activity and functionality of the proteins.

136

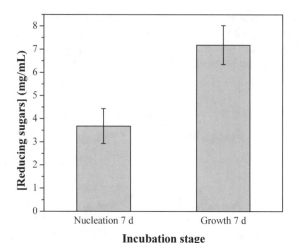

Incubation stage

Figure 4. Concentration of reducing sugars produced after 7 days of incubation when BSA and α-amylase were incorporated at the nucleation stage and growth stage.

It is, therefore, of extreme importance to evaluate the occurrence of conformational rearrangements on the protein structure after incorporation, which then may affect its activity/functionality. Furthermore, depending on the degree of conformational changes, different amino acids will be exposed at the surface, which will then control the interactions between the incorporated protein and the surrounding molecules or entities. Having into account the various interactions that can be established between the proteins, the substrate and the various ions present in solution during the nucleation and growth stages, the mineralization process in presence of biological molecules can be considered a rather complex subject. Nevertheless, the results of our work clearly show that Ca-P coatings, produced by a biomimetic technique, allow the efficient incorporation of α-amylase in the active form. This methodology may be used, for instances, as a way to control the degradation rate of starch-based biomaterials. Our results also suggest the possibility of modulating the properties of Ca-P coatings by incorporating different proteins during the coating preparation. More details on the incorporation of bioactive agents on biomimetic coatings can be found in references [93,94,119].

A better understanding of protein conformation on mineral surfaces may provide design principles useful for the modification of orthopaedic implants with coatings and biological growth factors that are aimed to enhance biocompatibility with surrounding tissue.

3. Which Kind of Functional Groups is Required to Render Biomaterials Self Mineralizable?

"It struck me that Nature's system must be a real beauty, because in chemistry we find that the associations are always in beautiful whole numbers -- there are no fractions."

Richard Buckminster Fuller (1895-1983)

Biological responses such as bone-bonding of the materials are very important in clinical applications. More specifically, a biologically active bone-like apatite layer has been shown to be a pre-requisite to bonding of artificial material to living bone [95,96]. As it as been discussed through this chapter a central idea in biomineralization is that the nucleation, growth and patterning of inorganic structures are controlled by interfacial interactions between mineral and protein/lipid surfaces. It is well known that these macromolecules contain functional groups, which are negatively charged at the crystallization pH [134]. The study of inorganic-organic interfaces is therefore an important aspect of biomimetic materials chemistry which seeks to develop new synthetic routes to functional materials that are organized on various length scales from the nano- to macroscopic.

Any argument surrounding mineral formation at interfaces would be deficient without consideration of nucleation and growth theory.

Heterogeneous nucleation does generally occur in biomineralization. It involves the formation of a new mineral phase (nuclei) on the surface of a substrate presents in the aqueous medium [3]. The formation of a *stable* nucleus requires certain energy since a new solid-liquid interface is created. At the same time some energy is released because of bonds formation inside inorganic clusters which are forming in the aqueous solution. Therefore, the free energy of formation of a nucleus is

$$\Delta G_N = \Delta G_I - \Delta G_B \qquad (1)$$

where ΔG_I is the energy necessary for new solid-liquid interface creation and ΔG_B is the energy released by bonds formation in the aggregate. The combination of those two functions goes through a free energy maximum ΔG_N^* which is known as activation energy for homogeneous nucleation and which is proportional to the interfacial free energy of nucleation (σ). The interfacial free energy of nucleation is composed of the contributions of the interfacial free energy of the crystal particle (σ_C), the substrate (σ_S) and the liquid (σ_L) phase:

$$\Delta G_N^* \sim \sigma_{CL} A_{CL} + (\sigma_{CS} - \sigma_{SL}) A_{CS} \qquad (2)$$

Where σ and A are respectively interfacial free energies and areas. The equation (2) shows that nucleation is favored if interaction between the forming nuclei and the substrate represent a net lower interfacial free energy than the particle/solution interfacial free energy [135]. This means that relatively small changes in σ_S can have a marked effect on nucleation rate. Of course the concentration of the solution and temperature are other key parameters which can control both nucleation and growth. For instances, SBF solution has a definite concentration and is one of the widely used in nowadays to test the bioactivity of different materials *in vitro* at 37 °C. However, the influence of these parameters is not going to be discussed in this chapter.

Interfacial energies can be lowered by the presence and/or modification of organic surfaces at the nucleation site [136]. Because the chemical bonding at the surface of mineral nuclei is primarily ionic, the organic interface must contain areas of high location **charge** where electrostatic, dipolar and hydrogen bonding interactions can take place during nucleation.

3.1. TAYLORING POLYMERIC SURFACES

Research in the biomimetic field is based upon the premise of surface functionalization [135]. As consequence, several strategies have been employed for creating functionalized interfaces that can be used to study the mechanisms controlling the heterogeneous nucleation. To reach such interface, in case of polymers can be achieved in two main ways [137] groups can be added by modification after polymerization or suitable monomers can be copolymerized with a conventional monomer. Which kind of functional groups/charges is required?

Since the electrostatic interaction triggers the initial step of nucleation, it is of primary interest to determine which ion Ca^{2+} or PO_4^{3-} adsorbs first on the surface. Tanahashi *et al.* [138] demonstrated that when a surface is negatively charged and then soaked in SBF acts as potent substrates for the apatite nucleation. They found that the incorporation of $-H_2PO_4$ and -COOH groups on self-assembled monolayers is effective for the apatite nucleation. On the other hand, functional groups like as $-CONH_2$, -OH, -NH_2 and $-CH_3$ had weaker apatite nucleating ability [138]. Among these functional groups, the incorporation of $-SO_3H$ groups onto polymer surfaces could also serve as the functional group for calcium phosphate nucleation [139,140]. These results proved that apatite formation occurs first by Ca^{2+} adsorption on and complexation with a negatively charged group of the material [138] (See Figure 5).

For example, Himeno *et al.* [141,142] confirmed these results by means of an electrophoresis study. They have shown that the surface potential by functional sites is initially negative, indicating that the initialization of apatite nucleation involves an electrostatic interaction between the functional sites and calcium ions. Therefore, functional groups able to become negatively charge at the blood plasma pH (≈ 7.4.) are assumed to be potentially effective for the apatite nucleation in an *in vivo* environment.

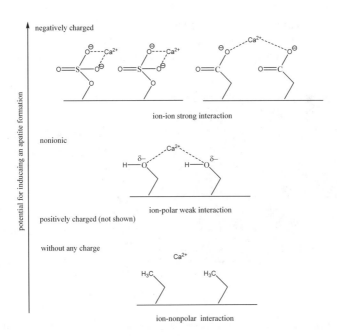

Figure 5. Quantitative ability of different functional groups to induce an apatite formation *in vitro*.

But the big question is: how can one design and prepare organized functionalized surfaces?

One of advantages of using polymers is the capacity to tailor their surface to achieve different properties such as making their surfaces more hydrophilic and capable of carrying functional groups.

For instances, one of the advantages of starch-based polymers is their ability to control their surface properties in such a way that a favourable interaction of the surface modified material and biological system is achieved [143]. It has been shown that SEVA-C, can associate a degradable behaviour with an interesting mechanical performance [89,144-147]. However, in terms of bone bonding, this polymer cannot induce the formation of an apatite layer without a bioactive coating or by using bioactive fillers as reported previously [108,109,113,148].

However, an organic, a quite hydrophilic material such as SEVA-C can be is a suitable model for apatite nucleation, as biological mineralization is thought to be induced by anionic functional groups. Therefore, the presence of reactive hydroxyl (-OH) groups on starch and vinyl alcohol justifies the present efforts in trying to incorporate another polar groups such as -COOH groups to obtain bioactive polymer. Our research group has been studying the effect of alkaline treatments to render starch based polymers self-mineralizable. For that purpose, it were used two different types of alkaline solutions: (i) calcium hydroxide solution ($Ca(OH)_2$) and (ii) sodium hydroxide solution (NaOH). Then, after the alkaline treatment for 24 hrs, the biodegradable polymeric substrate, SEVA-C, was immersed in SBF for several time periods up to 7 days.

The SEVA-C surface was completely covered with an apatite layer after the $Ca(OH)_2$ treatment and the subsequent soaking in SBF for 7 days (See the Fig. 6a). This layer was dense and compact. At higher magnifications these films evidenced a finer

140

structure, where needle-like crystals are agglomerated. Furthermore, the thin-film X-ray diffraction evidenced the diffraction maximal, which can be assigned to HA (JCPDS 9-432), confirming the formation of apatite layer on the surface of the polymer as it is shown in Figure 6b).

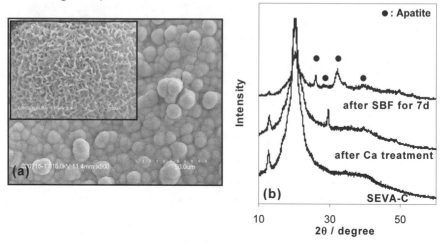

Figure 6. FE-SEM micrographs (a) and TF-XRD patterns (b) of the surface of Ca(OH)$_2$–treated SEVA-C after soaking in SBF for 7 days.

When SEVA-C was soaked in the NaOH solution and subsequently immersed in SBF the precipitation of apatite was observed only after 7 days (See the Figure 7a), which was confirmed by as it is shown in the Figure 7b.

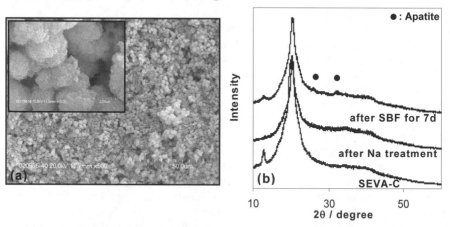

Figure 7. FE-SEM micrographs (a) and TF-XRD patterns (b) of the surface of NaOH–treated SEVA-C after soaking in SBF for 7 days.

As mentioned previously, the untreated SEVA-C cannot induce the formation of an apatite layer on its surface even after 7 days in SBF. This was confirmed by TF-XRD and Fourier Transformed Infra-Red with Attenuated Total Reflectance (FTIR-ATR) measurements (data not shown). The Ca(OH)$_2$ treatment was more efficient on the induction of bone-like apatite layer than a NaOH treatment. These results are attributed to the formation of -COOH groups (confirmed by FTIR-ATR measurements), and besides that, the water uptake capability of this polymer allowed the material to absorb higher quantities of Ca^{2+} ions from the calcium solution. As result, these groups together with Ca^{2+} ions, leads to an increase of the ionic activity product of the apatite and as result accelerate the apatite nucleation. In the case of using NaOH solution, the SEVA-C absorbed high quantities of Na^+ ion, and when soaked in SBF, these ions are released into the solution, which bound with the CO_3^- ion from SBF solution, and lead to the formation of sodium carbonate (Na_2CO_3). As a consequence, the formation of the apatite layer was delayed.

These results suggest that this rather simple treatment is a good method for surface functionalization and subsequent mineral nucleation and growth on biodegradable polymers to be used for bone related applications.

Very similar results were also obtained in our research group, with starch based blends after surface oxidation [149,150]. As mentioned above, starch itself contains many hydroxyl groups (nonionic, Fig. 5). In order to alternate nonionic starch hydroxyl groups with "loved" from the calcium ions negatively charged carboxyl groups, a surface oxidation by potassium permanganate/nitric acid system was performed [149]. Formation of an apatite layer was confirmed by FTIR-ATR, Induced Coupled Plasma (ICP), TF-XRD and SEM measurements on two different types of starch blends, (i) starch/cellulose acetate blends – SCA and (ii) starch/polycaprolactone blends – SPCL. After only 3 days of immersion in SBF only the formation of calcium phosphate nucleus was observed, but after 14 days it can been seen the formation of a dense film (Figure 8). Furthermore, TF-XRD analyses diffraction confirm the formation of an apatite layer on the surface of the polymer, as it can be seen in the Figure 9.

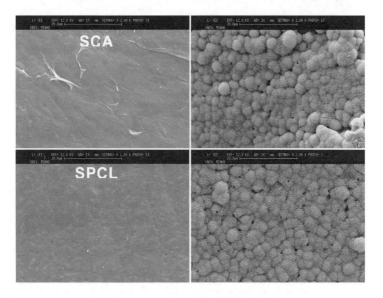

Figure 8. SEM micrographs of original and treated SCA and SPCL surfaces after 14 days immersing in SBF.

In the case of SEVA-C, the results obtained showed bioactivity only after being "embedded" with calcium ions by pretreatment with CaCl$_2$ solution. The same behaviour for –COOH (carboxymethylated chitin and gelatin gum) and –OH (curdlan gel) groups containing gels was reported by Kawashita *et al.* [151,152]. The hydroxyl containing gel was not able to form an apatite layer on its surface even after soaking in saturated Ca(OH)$_2$. In contrast, both carboxyl containing gels have capacity to induce the formation of an apatite layer after this pretreatment and subsequent immersion in SBF.

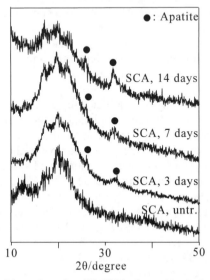

Figure 9. TF-XRD patterns of the surface of modified SCA after soaking in the SBF for different periods of time.

In fact, all those results are not surprising since they are in agreement with the model given to us by the Nature. According to this model [3,153], established by extraction of proteins from mineralized extracellular tissues such bone and shells, the organic matrix consists of a structural framework of hydrophobic macromolecules in association with **acidic macromolecules** (Figure 10) that act as a nucleation surface for biomineralization.

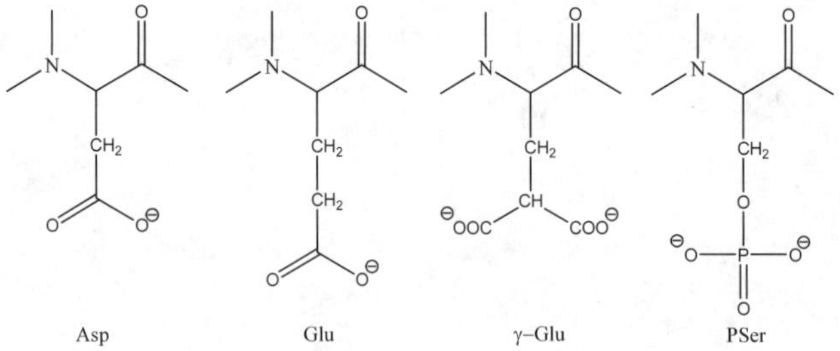

Figure 10. Some amino acid side chains associated with acidic macromolecules in biomineralization.

Of course, the presence of a negative charge is not the only requirement for formation of a bonelike apatite layer. As can be seen from Figure 5 calcium ions can only recognize those ion-binding sites if they have an appropriate space disposition. This geometric register [153] also "gives the face" of the crystal. For example, the spacing between aspartic acid residues are similar to the Ca-Ca distance in the nucleated (001) face of aragonite crystals from the nacreous layer of mollusk [154] shells. Very often, di- (oxalates, malonates, succinates, glutarates, maleates, etc.) [3] or polycarboxylate [155-157] are used as modifiers in the crystallization process. The choice of those modifiers and their concentration must be very careful since they can have an opposite effect. Kato *et al.* [157] have shown that increasing in the poly(acrylic acid) concentration results in lack of crystallization at the same time it is needed for the architecture of apatite crystals on modified pullulans [156]. Undesired bind of phosphocitrate [158] to hydroxyapatite can induce morphological changes or total cessation of apatite growing.

Nature is always using well-defined and high-organized structure as scaffolds. Unfortunately, unlike the self-assembly processes or chemisorption of functional molecules on a surface of a single crystal substrate, the surface modification of polymer substrates cannot always get highly organized surfaces [159] where the functional groups form uniform "lattices" across the whole substrate surfaces. Only small changes can result in different crystals lattice or in lack of nucleation. Just as an example it is worth stating that recently Sugawara *et al.* [156] showed that pullulan containing eight cholesteryl (hydrophobic, no charge) moieties per 100 glucose units gives $CaCO_3$ films with less regular structure than this one modified with three cholesteryl moieties when immersed in aqueous solution of $CaCl_2$ containing polyacrylic acid as an additive.

4. General Remarks and Future Perspectives

In summary, there is a large number of technologies (electronics, biotechnology, biosensors, material's fabrication such as ceramics, nanocomposites, coatings) which are using proteins, enzymes, peptides, functional groups to act as nucleators, anchoring units or growth modifiers for the synthesis of inorganic materials [26,73,98,156,160]. These approaches may be also very useful in many other applications, such as the development of novel materials for biomedical applications. Biomolecules, with different affinities for a certain polymeric material, may be anchored on its surface with favourable orientation in order to control the nucleating sites, growth kinetics and final structure of inorganic crystals. This may be achieved taking advantage of progresses done in the molecular design of recombinant proteins via genetic engineering proteins or by site-directed mutagenesis of existing proteins to obtain peptides or proteins that will bind specifically and selectively to inorganic surfaces [160]. In addition, the great diversity of enzymes available in Nature, associated with their unique and novel catalytic activities, may offer interesting opportunities to design new multifunction materials, as they are able to transform and control the surface and local chemistry.

There is no doubt that human beings have made possible efforts to create appropriate enhanced materials. As compared with their typical materials development strategies, the strategies of the Nature are based onto the following principles: (i) the weaker ones are victims of the stronger and (ii) trial and error approaches for a period of 100 millions years or more [161]. The first principle is also effective in the industrial society of man, but there is quite a difference in terms of the long-term trial and error experience. Therefore, if the experience from the natural world can be applied for the industrial materials, great advances in the field of materials research and also a big contribution to the society will be expected. There is still room for creating new structures or modify in a new way well known old ones, but always the scientists participating in this process must follow the way shown them by Nature, because the final word is from her, she will approve or reject what was created.

Acknowledgements

This work was partially supported by Portuguese Foundation for Science and Technology (FCT), through funds from the POCTI and/or FEDER programmes. I. B. Leonor, C. M. Alves and A. L. Oliveira thank the FCT for providing them PhD scholarships (SFRH/BD/9031/2002, SFRH/BD/11188/2002 and SFRH/BD/10956/2002) respectively. H. S. Azevedo and I. Pashkuleva thank the FCT for providing them post-doctoral scholarships (SFRH/BPD/5744/2001 and SFRH/BPD/8491/2002), respectively.

References

1. Heuer, A.H., Fink, D.J., Laraia, V.J., Arias, J.L., Calvert, P.D., Kendall, K., Messing, G.L., Blackwell, J., Rieke, P.C., Thompson, D.H. and et al. (1992) Innovative materials processing strategies: a biomimetic approach, *Science* **255**, 1098-1105.
2. Mann, S., Heywood, B.R., Rajam, S. and Birchall, J.D. (1988) Controlled Crystallization of $CaCO_3$ under Stearic-Acid Monolayers, *Nature* **334**, 692-695.
3. Mann, S. (2001) *Biomineralization, Principles and Concepts in Bioinorganic Materials Chemistry*, Oxford University Press, Oxford.
4. Mann, S. (1991) Biominerals - Flattery by Imitation, *Nature* **349**, 285-286.
5. Addadi, L. and Weiner, S. (1997) Biomineralization - A pavement of pearl, *Nature* **389**, 912-&.
6. Linde, A. and Goldberg, M. (1993) Dentinogenesis, *Crit Rev Oral Biol Med* **4**, 679-728.
7. Linde, A. (1989) Dentin matrix proteins: composition and possible functions in calcification, *Anat Rec* **224**, 154-166.
8. Zeichner-David, M. (2001) Is there more to enamel matrix proteins than biomineralization?, *Matrix Biology* **20**, 307-316.
9. Robinson, C., Weatherell, J.A. and Hohling, H.J. (1983) Formation and Mineralization of Dental Enamel, *Trends in Biochemical Sciences* **8**, 284-287.
10. Mann, S. (1997) The biomimetics of enamel: a paradigm for organized biomaterials synthesis, *Ciba Found Symp* **205**, 261-269; discussion 269-274.
11. Kampen, W.U., Claassen, H. and Kirsch, T. (1995) Mineralization and Osteogenesis in the Human First Rib Cartilage, *Annals of Anatomy-Anatomischer Anzeiger* **177**, 171-177.
12. Boskey, A.L., Stiner, D., Doty, S.B., Binderman, I. and Leboy, P. (1992) Studies of Mineralization in Tissue-Culture - Optimal Conditions for Cartilage Calcification, *Bone and Mineral* **16**, 11-36.
13. Weiner, S. and Wagner, H.D. (1998) The material bone: Structure mechanical function relations, *Annual Review of Materials Science* **28**, 271-298.
14. Anderson, H.C. (1984) Mineralization by matrix vesicles, *Scan Electron Microsc* 953-964.
15. Posner, A.S. (1985) The mineral of bone, *Clin Orthop* 87-99.
16. Mann, S. and Weiner, S. (1999) Biomineralization: structural questions at all length scales, *J Struct Biol* **126**, 179-181.
17. Mann, S. (2000) The chemistry of form, *Angewandte Chemie-International Edition* **39**, 3393-3406.
18. Mann, S., Archibald, D.D., Didymus, J.M., Douglas, T., Heywood, B.R., Meldrum, F.C. and Reeves, N.J. (1993) Crystallization at Inorganic-Organic Interfaces - Biominerals and Biomimetic Synthesis, *Science* **261**, 1286-1292.
19. Sarikaya, M., Fong, H., Frech, D.W. and Humbert, R. (1999) Biomimetic assembly of nanostructured materials, *Bioceramics* **293**, 83-97.

20. Murphy, W.L. and Mooney, D.J. (2001) Biomineralization via bioinspired variation in polymer surface chemistry., *Abstracts of Papers of the American Chemical Society* **222**, U344-U344.
21. Addadi, L., Berman, A., Moradianoldak, J. and Weiner, S. (1989) Protein-Crystal Interactions in Biomineralization, *Abstracts of Papers of the American Chemical Society* **197**, 115-Iaec.
22. Addadi, L., Berman, A., Moradianoldak, J. and Weiner, S. (1990) Tuning of Crystal Nucleation and Growth by Proteins - Molecular-Interactions at Solid-Liquid Interfaces in Biomineralization, *Croatica Chemica Acta* **63**, 539-544.
23. Addadi, L., Berman, A., Oldak, J.M. and Weiner, S. (1989) Structural and stereochemical relations between acidic macromolecules of organic matrices and crystals, *Connect Tissue Res* **21**, 127-134; discussion 135.
24. Addadi, L. and Weiner, S. (1986) Interactions between Acidic Macromolecules and Structured Crystal-Surfaces - Stereochemistry and Biomineralization, *Molecular Crystals and Liquid Crystals* **134**, 305-322.
25. Heywood, B.R. and Mann, S. (1994) Template-Directed Nucleation and Growth of Inorganic Materials, *Advanced Materials* **6**, 9-20.
26. Sarikaya, M., Tamerler, C., Jen, A.K., Schulten, K. and Baneyx, F. (2003) Molecular biomimetics: nanotechnology through biology, *Nature Materials* **2**, 577-585.
27. Calvert, P. (1994) Biomimetic mineralization: Processes and prospects, *Materials Science and Engineering: C* **1**, 69-74.
28. Weiner, S., Addadi, L. and Wagner, H.D. (2000) Materials design in biology, *Materials Science & Engineering C-Biomimetic and Supramolecular Systems* **11**, 1-8.
29. Weiner, S. and Addadi, L. (1997) Design strategies in mineralized biological materials, *Journal of Materials Chemistry* **7**, 689-702.
30. Ball, P. (2001) Life's lessons in design, *Nature* **409**, 413-416.
31. Liu, X.Y. and Lim, S.W. (2003) Templating and supersaturation-driven anti-templating: principles of biomineral architecture, *J Am Chem Soc* **125**, 888-895.
32. Posner, A.S. (1969) Crystal Chemistry of Bone Mineral, *Physiological Reviews* **49**, 760-&.
33. Jin, Y. and Yip, H.K. (2002) Supragingival calculus: Formation and control, *Critical Reviews in Oral Biology & Medicine* **13**, 426-441.
34. Tohda, H., Tsuchiya, Y., Kobayashi, T., Kishiro, H. and Yanagisawa, T. (1994) The Crystalline-Structure of Pancreatic Calculi, *Journal of Electron Microscopy* **43**, 57-61.
35. Rodan, G.A. (1992) Introduction to Bone Biology, *Bone* **13**, S3-S6.
36. Anderson, H.C. (2003) Matrix vesicles and calcification, *Curr Rheumatol Rep* **5**, 222-226.
37. Anderson, H.C., Sipe, J.B., Hessle, L., Dhamyamraju, R., Atti, E., Camacho, N.P. and Millan, J.L. (2004) Impaired calcification around matrix vesicles of growth plate and bone in alkaline phosphatase-deficient mice, *American Journal of Pathology* **164**, 841-847.
38. Anderson, H.C. (1995) Molecular Biology of Matrix Vesicles, *Clin Orthop* **314**, 266-280.
39. Anderson, H.C. (1981) Normal and abnormal mineralization in mammals, *Trans Am Soc Artif Intern Organs* **27**, 702-708.
40. Kirsch, T., Nah, H.D. and Pacifici, M. (1998) The role of collagen-annexin V interactions in mineralization of skeletal tissues, *Matrix Biology* **17**, 162-162.
41. Gillette, J.M. and Nielsen-Preiss, S.M. (2004) The role of annexin 2 in osteoblastic mineralization, *Journal of Cell Science* **117**, 441-449.
42. Blakemore, R. (1975) Magnetotactic bacteria, *Science* **190**, 377-379.
43. Eanes, E.D. (1992) Mixed phospholipid liposome calcification, *Bone and Mineral* **17**, 269-272.
44. Sikavitsas, V.I., Temenoff, J.S. and Mikos, A.G. (2001) Biomaterials and bone mechanotransduction, *Biomaterials* **22**, 2581-2593.
45. van Leeuwen, J.P.T.M., van Driel, M., van den Bemd, G.J.C.M. and Pols, H.A.P. (2001) Vitamin D control of osteoblast function and bone extracellular matrix mineralization, *Critical Reviews in Eukaryotic Gene Expression* **11**, 199-226.
46. Gorski, J.P. (1992) Acidic phosphoproteins from bone matrix: a structural rationalization of their role in biomineralization, *Calcif Tissue Int* **50**, 391-396.
47. Ganss, B., Kim, R.H. and Sodek, J. (1999) Bone sialoprotein, *Crit Rev Oral Biol Med* **10**, 79-98.
48. MacNeil, R.L., Berry, J., D'Errico, J., Strayhorn, C., Piotrowski, B. and Somerman, M.J. (1995) Role of two mineral-associated adhesion molecules, osteopontin and bone sialoprotein, during cementogenesis, *Connect Tissue Res* **33**, 1-7.
49. Engfeldt, B. and Hjerpe, A. (1976) Glycosaminoglycans and proteoglycans of human bone tissue at different stages of mineralization, *Acta Pathol Microbiol Scand [A]* **84**, 95-106.
50. Hoffmann, A. and Gross, G. (2001) BMP signaling pathways in cartilage and bone formation, *Crit Rev Eukaryot Gene Expr* **11**, 23-45.

146

51. Li, R.H. and Wozney, J.M. (2001) Delivering on the promise of bone morphogenetic proteins, *Trends in Biotechnology* **19**, 255-265.
52. Yakar, S. and Rosen, C.J. (2003) From mouse to man: redefining the role of insulin-like growth factor-I in the acquisition of bone mass, *Exp Biol Med (Maywood)* **228**, 245-252.
53. Weiss, R.E., Itatani, C., Marshall, G.J. and Nimni, M.E. (1981) The role of fibronectin and substrate size in attachment of rat bone marrow cells to an osteoinductive matrix, *Scan Electron Microsc* **4**, 183-188.
54. Weiss, R.E. and Reddi, A.H. (1980) Synthesis and localization of fibronectin during collagenous matrix-mesenchymal cell interaction and differentiation of cartilage and bone in vivo, *Proc Natl Acad Sci U S A* **77**, 2074-2078.
55. Weiss, R.E. and Reddi, A.H. (1981) Appearance of fibronectin during the differentiation of cartilage, bone, and bone marrow, *Journal of Cell Biology* **88**, 630-636.
56. Rodan, G.A. and Harada, S. (1997) The missing bone, *Cell* **89**, 677-680.
57. Dunstan, C.R., Boyce, R., Boyce, B.F., Garrett, I.R., Izbicka, E., Burgess, W.H. and Mundy, G.R. (1999) Systemic administration of acidic fibroblast growth factor (FGF-1) prevents bone loss and increases new bone formation in ovariectomized rats, *Journal of Bone and Mineral Research* **14**, 953-959.
58. Rodan, G.A. and Martin, T.J. (2000) Therapeutic approaches to bone diseases, *Science* **289**, 1508-1514.
59. Rodan, G.A. (1991) Mechanical Loading, Estrogen Deficiency, and the Coupling of Bone-Formation to Bone-Resorption, *Journal of Bone and Mineral Research* **6**, 527-530.
60. Rodan, G.A. and Hirsch, L.J. (1996) Treatment of bone in elderly subjects: Calcium, vitamin D, fluor, bisphosphonates, calcitonin, *Hormone Research* **45**, 300-301.
61. Cowles, E.A., DeRome, M.E., Pastizzo, G., Brailey, L.L. and Gronowicz, G.A. (1998) Mineralization and the expression of matrix proteins during in vivo bone development, *Calcified Tissue International* **62**, 74-82.
62. Roach, H.I. (1994) Why Does Bone-Matrix Contain Noncollagenous Proteins - the Possible Roles of Osteocalcin, Osteonectin, Osteopontin and Bone Sialoprotein in Bone Mineralization and Resorption, *Cell Biology International* **18**, 617-628.
63. Pacifici, M., Oshima, O., Fisher, L.W., Young, M.F., Shapiro, I.M. and Leboy, P.S. (1990) Changes in osteonectin distribution and levels are associated with mineralization of the chicken tibial growth cartilage, *Calcif Tissue Int* **47**, 51-61.
64. Robey, P.G. (1989) The Biochemistry of Bone, *Endocrinology and Metabolism Clinics of North America* **18**, 859-902.
65. Canalis, E., Varghese, S., Mccarthy, T.L. and Centrella, M. (1992) Role of Platelet Derived Growth-Factor in Bone Cell-Function, *Growth Regulation* **2**, 151-155.
66. Okamoto, Y. and Hidaka, S. (1994) Studies on Calcium-Phosphate Precipitation - Effects of Metal-Ions Used in Dental Materials, *Journal of Biomedical Materials Research* **28**, 1403-1410.
67. Stupp, S.I. and Braun, P.V. (1997) Molecular manipulation of microstructures: biomaterials, ceramics, and semiconductors, *Science* **277**, 1242-1248.
68. Walsh, D., Kingston, J.L., Heywood, B.R. and Mann, S. (1993) Influence of Monosaccharides and Related Molecules on the Morphology of Hydroxyapatite, *Journal of Crystal Growth* **133**, 1-12.
69. Sarikaya, M. (1999) Biomimetics: Materials fabrication through biology, *Proceedings of the National Academy of Sciences of the United States of America* **96**, 14183-14185.
70. Douglas, T. (2003) Materials science. A bright bio-inspired future, *Science* **299**, 1192-1193.
71. Rodan, G.A., Rodan, S.B., Yeh, C.K. and Thompson, D.D. (1986) Prostaglandin and Bone Remodeling, *Journal of Cellular Biochemistry* 107-107.
72. Chien, H.H., Lin, W.L. and Cho, M.I. (1999) Interleukin-1 beta-induced release of matrix proteins into culture media causes inhibition of mineralization of nodules formed by periodontal ligament cells in vitro, *Calcified Tissue International* **64**, 402-413.
73. Lakshminarayanan, R., Kini, R.M. and Valiyaveettil, S. (2002) Investigation of the role of ansocalcin in the biomineralization in goose eggshell matrix, *Proc Natl Acad Sci U S A* **99**, 5155-5159.
74. Aizenberg, J., Muller, D.A., Grazul, J.L. and Hamann, D.R. (2003) Direct fabrication of large micropatterned single crystals, *Science* **299**, 1205-1208.
75. Aizenberg, J., Black, A.J. and Whitesides, G.M. (1999) Control of crystal nucleation by patterned self-assembled monolayers, *Nature* **398**, 495-498.
76. Liu, Y., Layrolle, P., de Bruijn, J., van Blitterswijk, C. and de Groot, K. (2001) Biomimetic coprecipitation of calcium phosphate and bovine serum albumin on titanium alloy, *Journal of Biomedical Materials Research* **57**, 327-335.
77. Zeng, H., Chittur, K.K. and Lacefield, W.R. (1999) Analysis of bovine serum albumin adsorption on calcium phosphate and titanium surfaces, *Biomaterials* **20**, 377-384.

78. Liu, Y., Stigter, M., Groot, K.d. and Layrolle, P. (2002) Protein modulate the properties of biomimetic calcium phosphate coatings of titanium implants, *Engineering Materials* **218-220**, 157-160.
79. Liu, Y., Hunziker, E.B., Randall, N.X., de Groot, K. and Layrolle, P. (2003) Proteins incorporated into biomimetically prepared calcium phosphate coatings modulate their mechanical strength and dissolution rate, *Biomaterials* **24**, 65-70.
80. Addadi, L. and Weiner, S. (1992) Control and Design Principles in Biological Mineralization, *Angewandte Chemie-International Edition in English* **31**, 153-169.
81. Green, D., Walsh, D., Mann, S. and Oreffo, R.O. (2002) The potential of biomimesis in bone tissue engineering: lessons from the design and synthesis of invertebrate skeletons, *Bone* **30**, 810-815.
82. Atlan, G., Balmain, N., Berland, S., Vidal, B. and Lopez, E. (1997) Reconstruction of human maxillary defects with nacre powder: histological evidence for bone regeneration, *Comptes Rendus De L Academie Des Sciences Serie Iii-Sciences De La Vie-Life Sciences* **320**, 253-258.
83. Addadi, L. and Weiner, S. (1985) Interactions between Acidic Proteins and Crystals - Stereochemical Requirements in Biomineralization, *Proceedings of the National Academy of Sciences of the United States of America* **82**, 4110-4114.
84. Fujisawa, R., Wada, Y., Nodasaka, Y. and Kuboki, Y. (1996) Acidic amino acid-rich sequences as binding sites of osteonectin to hydroxyapatite crystals, *Biochimica Et Biophysica Acta-Protein Structure and Molecular Enzymology* **1292**, 53-60.
85. Addadi, L., Maroudas, N.G., Shay, E. and Weiner, S. (1985) Calcite Nucleation on Model Polymers - a Cooperative Hypothesis, *Bone* **6**, 483-483.
86. Malafaya, P.B., Silva, G.A., Baran, E.T. and Reis, R.L. (2002) Drug delivery therapies I - General trends and its importance on bone tissue engineering applications, *Current Opinion in Solid State & Materials Science* **6**, 283-295.
87. Malafaya, P.B., Silva, G.A., Baran, E.T. and Reis, R.L. (2002) Drug delivery therapies II. Strategies for delivering bone regenerating factors, *Current Opinion in Solid State & Materials Science* **6**, 297-312.
88. Reis, R.L. and Cunha, A.M. (1995) Characterization of two biodegradable polymers of potential application within the biomaterials field, *Journal of Materials Science-Materials in Medicine* **6**, 786-792.
89. Reis, R.L., Cunha, A.M., Allan, P.S. and Bevis, M.J. (1996) Mechanical behavior of injection-molded starch-based polymers, *Polymers for Advanced Technologies* **7**, 784-790.
90. Mendes, S.C., Bezemer, J., Claase, M.B., Grijpma, D.W., Bellia, G., Degli-Innocenti, F., Reis, R.L., De Groot, K., Van Blitterswijk, C.A. and De Bruijn, J.D. (2003) Evaluation of two biodegradable polymeric systems as substrates for bone tissue engineering, *Tissue Engineering* **9**, S91-S101.
91. Gomes, M.E., Reis, R.L., Cunha, A.M., Blitterswijk, C.A. and de Bruijn, J.D. (2001) Cytocompatibility and response of osteoblastic-like cells to starch-based polymers: effect of several additives and processing conditions, *Biomaterials* **22**, 1911-1917.
92. Marques, A.P., Reis, R.L. and Hunt, J.A. (2002) The biocompatibility of novel starch-based polymers and composites: in vitro studies, *Biomaterials* **23**, 1471-1478.
93. Leonor, I.B., Azevedo, H.S., Alves, C.M. and Reis, R.L. (2003) Effects of the incorporation of proteins and active enzymes on biomimetic calcium-phosphate coatings, in B. Ben-Nissan, D. Sher and W. Walsh (eds.), *Bioceramics 15*, Trans Tech Publications, Zurich, pp. 97-100.
94. Azevedo, H.S., Leonor, I.B., Alves, C.M., Goldsmith, R.J. and Reis, R.L. (2004) Influence of Protein Incorporation in the Nucleation and Growth of Biomimetic Calcium Phosphate Coatings, *7th World Biomaterials Congress*, Sydney.
95. Kokubo, T., Ito, S., Huang, Z.T., Hayashi, T., Sakka, S., Kitsugi, T. and Yamamuro, T. (1990) Ca,P-rich layer formed on high-strength bioactive glass-ceramic A-W, *Journal of Biomedical Materials Research* **24**, 331-343.
96. Reis, R.L., Cunha, A.M., Fernandes, M.H. and Correia, R.N. (1997) Treatments to induce the nucleation and growth of apatite-like layers on polymeric surfaces and foams, *Journal of Materials Science-Materials in Medicine* **8**, 897-905.
97. Abe, Y., Kokubo, T. and Yamamuro, T. (1990) Apatite coating on ceramics, metals and polymers utilizing a biological process, *Journal of Materials Science: Materials in Medicine* **1**, 233-238.
98. Boskey, A.L. (1998) Will biomimetics provide new answers for old problems of calcified tissues?, *Calcif Tissue Int* **63**, 179-182.
99. Stupp, S.I., LeBonheur, V.V., Walker, K., Li, L.S., Huggins, K.E., Keser, M. and Amstutz, A. (1997) Supramolecular Materials: Self-Organized Nanostructures, *Science* **276**, 384-389.
100. Kokubo, T., Kim, H.M., Kawashita, M. and Nakamura, T. (2001) Process of calcification on artificial materials, *Z Kardiol* **90 Suppl 3**, 86-91.

148

101. Kokubo, T., Kim, H.M., Kawashita, M., Takadama, H., Miyazaki, T., Uchida, M. and Nakamura, T. (2000) Nucleation and growth of apatite an amorphous phases in simulated body fluid, *Glass Science and Technology-Glastechnische Berichte* **73**, 247-254.
102. Kokubo, T., Kim, H.M., Miyaji, F., Takadama, H. and Miyazaki, T. (1999) Ceramic-metal and ceramic-polymer composites prepared by a biomimetic process, *Composites Part a-Applied Science and Manufacturing* **30**, 405-409.
103. Kokubo, T. (1996) Formation of biologically active bone-like apatite on metals and polymers by a biomimetic process, *Thermochimica Acta* **280-281**, 479-490.
104. Hata, K., Kokubo, T., Nakamura, T. and Yamamuro, T. (1995) Growth of a Bonelike Apatite Layer on a Substrate by a Biomimetic Process, *Journal of the American Ceramic Society* **78**, 1049-1053.
105. Kokubo, T., Kushitani, H., Sakka, S., Kitsugi, T. and Yamamuro, T. (1990) Solutions able to reproduce in vivo surface-structure changes in bioactive glass-ceramic A-W, *Journal of Biomedical Materials Research* **24**, 721-734.
106. Oliveira, A.L., Mano, J.F. and Reis, R.L. (2003) Nature-inspired calcium phosphate coatings: present status and novel advances in the science of mimicry, *Current Opinion in Solid State and Materials Science* **7**, 309-318.
107. Miyaji, F., Kim, H.M., Handa, S., Kokubo, T. and Nakamura, T. (1999) Bonelike apatite coating on organic polymers: novel nucleation process using sodium silicate solution, *Biomaterials* **20**, 913-919.
108. Oliveira, A.L., Elvira, C., Reis, R.L., Vásquez, B. and Román, J.S. (1999) Surface modification tailors the characteristics of biomimetic coatings nucleated on starch-based polymers, *Journal of Materials Science: Materials in Medicine* **10**, 827-835.
109. Oliveira, A.L., Malafaya, P.B. and Reis, R.L. (2003) Sodium silicate gel as a precursor for the in vitro nucleation and growth of a bone-like apatite coating in compact and porous polymeric structures, *Biomaterials* **24**, 2575-2584.
110. Oliveira, A.L., Alves, C.M. and Reis, R.L. (2002) Cell adhesion and proliferation on biomimetic calcium-phosphate coatings produced by a sodium silicate gel methodology, *Journal of Materials Science-Materials in Medicine* **13**, 1181-1188.
111. Oliveira, A.L., Gomes, M.E., Malafaya, P.B. and Reis, R.L. (2003) Biomimetic coating of starch based polymeric foams produced by a calcium silicate based methodology, in B. Ben-Nissan, D. Sher and W. Walsh (eds.), *Bioceramics 15*, Trans Tech Publications, Zurich, pp. 101-104.
112. Oliveira, A.L., Leonor, I.B., Elvira, C., Azevedo, M.C., Pashkuleva, I. and Reis, R.L. (2002) Surface Treatments and Pre-calcification Routes to Enhance Cell Adhesion and Proliferation, in R.L.R.a.D. Cohn (eds.), *Polymer Based Systems on Tissue Engineering, Replacement and Regeneration*, Kluwer Press, Drodercht, pp. 183-217.
113. Leonor, I.B. and Reis, R.L. (2003) An innovative auto-catalytic deposition route to produce calcium-phosphate coatings on polymeric biomaterials, *Journal of Materials Science-Materials in Medicine* **14**, 435-441.
114. Doheny, J.G., Jervis, E.J., Guarna, M.M., Humphries, R.K., Warren, R.A.J. and Kilburn, D.G. (1999) Cellulose as an inert matrix for presenting cytokines to target cells: production and properties of a stem cell factor-cellulose-binding domain fusion protein, *Biochemical Journal* **339**, 429-434.
115. Jessel, N., Atalar, F., Lavalle, P., Mutterer, J., Decher, G., Schaaf, P., Voegel, J.C. and Ogier, J. (2003) Bioactive coatings based on a polyelectrolyte multilayer architecture functionalized by embedded proteins, *Advanced Materials* **15**, 692-695.
116. Puleo, D.A., Kissling, R.A. and Sheu, M.S. (2002) A technique to immobilize bioactive proteins, including bone morphogenetic protein-4 (BMP-4), on titanium alloy, *Biomaterials* **23**, 2079-2087.
117. Costa, S.A., Azevedo, H.S. and Reis, R.L. (2004) Enzyme Immobilization in Biodegradable Polymers for Biomedical Applications, in R.L. Reis and J.S. Román (eds.), *Biodegradable Systems in Tissue Engineering and Regenerative Medicine*, CRC Press, Boca Raton, *in press*.
118. Puleo, D.A. and Nanci, A. (1999) Understanding and controlling the bone-implant interface, *Biomaterials* **20**, 2311-2321.
119. Leonor, I.B., Azevedo, H.S., Alves, C.M. and Reis, R.L. (2004) Biomimetic Coatings, Proteins and Biocatalysts: a New Approach to Tailor the Properties of Biodegradable Polymers, in R.L. Reis and J.S. Román (eds.), *Biodegradable Systems in Tissue Engineering and Regenerative Medicine*, CRC Press, Boca Raton, *in press*.
120. Martin, R.I. and Brown, P.W. (1994) Formation of Hydroxyapatite in Serum, *Journal of Materials Science-Materials in Medicine* **5**, 96-102.
121. Klinger, A., Steinberg, D., Kohavi, D. and Sela, M.N. (1997) Mechanism of adsorption of human albumin to titanium in vitro, *Journal of Biomedical Materials Research* **36**, 387-392.
122. Combes, C., Rey, C. and Freche, M. (1999) In vitro crystallization of octacalcium phosphate on type I collagen: influence of serum albumin, *Journal of Materials Science-Materials in Medicine* **10**, 153-160.

123. Azevedo, H.S., Gama, F.M. and Reis, R.L. (2003) In vitro assessment of the enzymatic degradation of several starch based biomaterials, *Biomacromolecules* **4**, 1703-1712.
124. Boskey, A.L. (1998) Biomineralization: conflicts, challenges, and opportunities, *J Cell Biochem Suppl* **30-31**, 83-91.
125. Cuisinier, F.J.G. (1996) Bone mineralization, *Current Opinion in Solid State and Materials Science* **1**, 436-439.
126. Combes, C. and Rey, C. (2002) Adsorption of proteins and calcium phosphate materials bioactivity, *Biomaterials* **23**, 2817-2823.
127. Couchourel, D., Escoffier, C., Rohanizadeh, R., Bohic, S., Daculsi, G., Fortun, Y. and Padrines, M. (1999) Effects of fibronectin on hydroxyapatite formation, *J Inorg Biochem* **73**, 129-136.
128. Eiden-Aβmann, S., Viertelhaus, M., Heiss, A., Hoetzer, K.A. and Felsche, J. (2002) The influence of amino acids on the biomineralization of hydroxyapatite in gelatin, *J Inorg Biochem* **91**, 481-486.
129. Fujisawa, R. and Kuboki, Y. (1991) Preferential adsorption of dentin and bone acidic proteins on the (100) face of hydroxyapatite crystals, *Biochimica Et Biophysica Acta* **1075**, 56-60.
130. Hunter, G.K., Hauschka, P.V., Poole, A.R., Rosenberg, L.C. and Goldberg, H.A. (1996) Nucleation and inhibition of hydroxyapatite formation by mineralized tissue proteins, *Biochemical Journal* **317** (**Pt 1**), 59-64.
131. Campbell, A.A. and Nancollas, G.H. (1991) The mineralization of calcium phosphate on separated salivary protein films, *Colloids and Surfaces* **54**, 33-40.
132. Marques, P.A., Serro, A.P., Saramago, B.J., Fernandes, A.C., Magalhaes, M.C. and Correia, R.N. (2003) Mineralisation of two phosphate ceramics in HBSS: role of albumin, *Biomaterials* **24**, 451-460.
133. Reis, R.L. and Cunha, A.M. (2001) Starch and Starch Based Thermoplastics, in K.H.J. Buschow, R.W. Cahn, M.C. Flemings, B. Ilschner, E.J. Kramer and S. Mahajan (eds.), *Biological and Biomimetic Materials*, Pergamon - Elsevier Science, Amsterdam, pp. 8810-8816.
134. Weiner, S. (1986) Organization of extracellularly mineralized tissues: A comparative study of biological crystal growth, *CRC Crit. Rev. Biochem.* **20**, 365-408.
135. Campbell, A.A. (1999) Interfacial regulation of crystallization in aqueous environments, *Current Opinion in Colloid & Interface Science* **4**, 40-45.
136. Mann, S. (1988) Molecular recognition in biomineralization, *Nature* **332**, 119-124.
137. Cunliffe, D., Pennadam, S., Alexander, C. (2004) Synthetic and biological polymers - merging the interface, *European Polymer Journal* **40**, 5-25.
138. Tanahashi, M. and Matsuda, T. (1997) Surface functional group dependence on apatite formation on self-assembled monolayers in a simulated body fluid, *Journal of Biomedical Materials Research* **34**, 305-315.
139. Leonor, I.B., Balas, F., Kim, H.-M., Kokubo, T. and Reis, R.L. (2003) Apatite-forming ability of polymers with SO₃H groups in SBF, *18th Meeting of the European Society For Biomaterials*, Stuttgart, Germany, T027.
140. Leonor, I.B., Balas, F., Kim, H.-M., Kokubo, T. and Reis, R.L. (2003) Ability of EVOH Polymeric Biomaterials Modified with Sulfonic Functional Groups, *29th Annual Meeting of the Society For Biomaterials*, Reno, Nevada, USA, 387.
141. Himeno, T., kim, H.-M., Kaneko, H., Kawashita, M., Kokubo, T. and Nakamura, T. (2002) Surface structural changes of sintered hydroxyapatite in terms of surface charge, in B. Ben-Nissan, D. Sher and W. Walsh (eds.), *Bioceramics Vol. 15*, Trans Tech Publications, Switzerland, pp. 457-460.
142. Kim, H.M., Himeno, T., Kawashita, M., Lee, J.H., Kokubo, T. and Nakamura, T. (2003) Surface potential change in bioactive titanium metal during the process of apatite formation in simulated body fluid, *Journal of Biomedical Materials Research* **67A**, 1305-1309.
143. Demirgoz, D., Elvira, C., Mano, J.F., Cunha, A.M., Piskin, E. and Reis, R.L. (2000) Chemical modification of starch based biodegradable polymeric blends: effects on water uptake, degradation behaviour and mechanical properties, *Polymer Degradation and Stability* **70**, 161-170.
144. Sousa, R.A., Mano, J.F., Reis, R.L., Cunha, A.M. and Bevis, M.J. (2002) Mechanical performance of starch based bioactive composite biomaterials molded with preferred orientation, *Polymer Engineering and Science* **42**, 1032-1045.
145. Reis, R.L., Mendes, S.C., Cunha, A.M. and Bevis, M.J. (1997) Processing and in vitro degradation of starch/EVOH thermoplastic blends, *Polymer International* **43**, 347-352.
146. Vaz, C.M., Reis, R.L. and Cunha, A.M. (2001) Degradation model of starch-EVOH plus HA composites, *Materials Research Innovations* **4**, 375-380.
147. Altpeter, H., Bevis, M.J., Gomes, M.E., Cunha, A.M. and Reis, R.L. (2003) Shear controlled orientation in injection moulding of starch based blends intended for medical applications, *Plastics Rubber and Composites* **32**, 173-181.

148. Leonor, I.B., Ito, A., Onuma, K., Kanzaki, N. and Reis, R.L. (2003) In vitro bioactivity of starch thermoplastic/hydroxyapatite composite biomaterials: an in situ study using atomic force microscopy, *Biomaterials* **24**, 579-585.

149. Pashkuleva, I., Marques, A., Vaz, F. and Reis, R.L. (2004) Surface modification of starch based blends using potassium permanganate-nitric acid system and its effect on the adhesion and proliferation of osteoblast-like cells, *J. Mat. Sci.: Mat. in Med.* **Accepted**.

150. Pashkuleva, I., Marques, A., Vaz, F. and Reis, R.L. (2003) Surface modification of starch based biomaterials can simultaneously enhance cell adhesion and proliferation and induce bioactivity, *18th conference of European Society of Biomaterials, Sttutgart, Germany,* **T103**.

151. Kawashita, M., Nakao, M., Minoda, M., Kim, H.M., Beppu, T., Miyamoto, T., Kokubo, T. and Nakamura, T. (2003) Apatite-forming ability of carboxyl group-containing polymer gels in a simulated body fluid, *Biomaterials* **24**, 2477-2484.

152. Kokubo, T., Hanakawa, M., Kawashita, M., Minoda, M., Beppu, T., Miyamoto, T., and Nakamura, T. (2004) Apatite formation on non-woven fabric of carboxymethylated chitin in SBF, *Biomaterials* **25**, 4485-4488.

153. Mann, S. (1993) Molecular tectonics in biomineralization and biomimetic materials chemistry, *Nature* **365**, 499-505.

154. Weiner, S., Traub, W. (1984) Macromolecules in mollusk shells and their functions in biomineralization, *Phil. Trans. R. Soc.* **B304**, 425-434.

155. Iwatsubo, T., Sumaru, K., Kanamori, T., Yamaguchi, T., Sinbo, T. (2004) Preferential mineralization of $CaCO_3$ layers on polymer surfaces from $CaCl_2$ and water-soluble carbonate salt solutions superstaurated by poly(acrylic acid), *J. Appl. Pol. Sci.* **91**, 3627-3634.

156. Sugawara, A., Ishii, T., Kato, T. (2003) Self-organized calcium carbonate with regular relief structures, *Angew. Chem. Int. Ed.* **42**, 5299-5303.

157. Kato, T., Suzuki, T., Amamiya, T., Irie, T., Komiyama, M., Yui, H. (1998) Effects of macromolecules on the crystallization of CaCO3 the formation of organic/inorganic composites, *Supramol. Sci.* **5**, 411-415.

158. Wierzbicki, A., Cheung, H.S. (2000) Molecular modeling of inhibition of hydroxyapatite by hposphocitrate, *J. Mol. Str. (Theochem)* **529**, 73-82.

159. Mao, C., Li, H., Cui, F., Ma, Ch., Feng, Q. (1999) Oriented growth of phosphates on polycrystalline titanium in a process mimicking biomineralization, *Journal of Crystal Growth* **206**, 308-321.

160. Shiba, K., Honma, T., Minamisawa, T., Nishiguchi, K. and Noda, T. (2003) Distinct macroscopic structures developed from solutions of chemical compounds and periodic proteins, *Embo Reports* **4**, 148-153.

161. Asai, S., Koumoto, K., Matsushita, Y., Yashima, E., Morinaga, M., Takeda, K., Iritani, E., Tagawa, T., Tanahashi, M., Miyazawa, K. (2003) Advances in nature-guided materials processing, *Sci. Techn. Adv. Mat. in press.*

LEARNING FROM MARINE CREATURES HOW TO DESIGN MICRO-LENSES

J. AIZENBERG
Bell Laboratories/Lucent Technologies
600 Mountain Ave., Murray Hill, NJ 07974, USA

G. HENDLER
Natural History Museum of Los Angeles County
Los Angeles, CA 90007, USA

1. Introduction

The increasing technological requirement for a new generation of optical devices with novel architectures, tunability and tailored properties, provides a potent stimulus to the academic study of optical systems in living organisms. Mimicking Nature's methods of biological manufacture is proving to be a major step forward in modern technology [1-7]. In general, a successful bio-inspired engineering effort includes three key objectives: (i) To search for smart biological solutions in design, synthesis, and integration of complex materials and systems; (ii) To learn about their structure, properties, tunability and mechanisms of formation; (iii) To identify new bio-inspired synthetic strategies and to apply this knowledge to the fabrication of novel, superior materials and devices.

We are interested in discovering natural optical systems, whose hierarchical architecture and hybrid character offer outstanding optical properties and enable multi-faceted roles for these units. Biology provides a multitude of varied, new paradigms for the development of adaptive optical networks. Recently, we have shown that biologically formed optical systems are often unique in their ability to perform multiple functions – optical and structural. For example, the study of light-sensitive ophiuroids showed the skeletal dorsal arm plates of some species not only afford protection [8,9], but also form highly efficient micro-lens arrays [10]. Other research has concentrated on the exceptional fiber-optical properties of siliceous spicules of certain hexactinellid sponges, whose primary function is structural (either skeletal or anchoring) [11-13].

This review describes the design and optical features of the natural ophiuroid micro-lenses, summarizes their advantageous properties, and shows our first, bio-inspired engineering efforts in the fabrication of new optical structures.

R.L. Reis and S. Weiner (eds.),
Learning from Nature How to Design New Implantable Biomaterials, 151-166.
© 2004 *Kluwer Academic Publishers. Printed in the Netherlands.*

2. Structural and Optical Characterization of Biologically Formed Micro-Lenses

Echinodermata, the group of animals including sea stars and sea urchins, show various levels of photosensitivity. An especially interesting example is presented by brittlestars (Ophiuroidea) in the genus *Ophiocoma*, which exhibit strong light responses. It has been generally believed that many echinoderm's have an extraocular sensitivity to light that is supported by unspecialized, "diffuse," or dermal receptors [14-16]. The absence of specialized "eyes" is not, however, readily reconciled with such reactions as the color-change and negative phototactic behavior observed in certain species. The variations in color and escape mechanisms are most striking in a brittlestar *Ophiocoma wendtii* (Fig. 1) [17, 18]. The individuals are a dark brown color during the day, and are strikingly banded with gray and black during the night. It has been documented that *O. wendtii* is able to escape from predatory fish by hiding in coral crevices, and that it may be able to sense shelter at a distance [18,19]. Recently, we suggested that these strong responses to light might be related to the presence of a characteristic lensar extension of the brittlestar skeleton, which may constitute an advanced photoreception system [10,17,18].

Figure 1. The same individual of the brittlestar *O. wendtii*, photographed during the day (top) and during the night (bottom).

Echinoderms use calcium carbonate for their skeleton construction. Each skeletal element (ossicle) is a single crystal of calcite [8,9,20,21]. Unlike biogenic calcite of other phyla or abiologically formed calcite crystals, echinoderm skeletons are composed of a unique, intricately shaped, three-dimensional, single-crystalline meshwork with smooth and curved surfaces lacking crystal faces (so-called, stereom) (Fig. 2). In the living animal, the calcite meshwork is filled with soft tissue (so-called, stroma), and the ossicles are

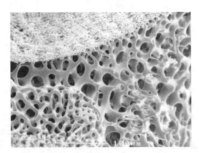

Figure 2. Typical stereom structure in echinoderm skeletons. The entire elaborate mesh is a single calcite crystal.

attached to one and other by connective tissue and muscle. Every joint of a brittlestar arm is composed of five major ossicles, including two lateral arm plates which support several spines and tentacle scales , one dorsal and one ventral arm plate. These surround the arm and enclose a large internal skeletal ossicle (vertebra) that is adapted for articulation. The analysis of the ultrastructure of these skeletal units showed that while the stereom in the vertebrae, spines, tentacle scales, and ventral arm plates generally share a typical design (Fig. 2), the dorsal arm plates (DAP) and portions of the lateral arm plates in certain species of *Ophiocoma* exhibit an unusual extension of the stereom – a thick, transparent, layer (~ 40 μm thick) that covers the external surface of the plate [10,17]. It is composed of a close-set array of hemispherical calcitic structures (reaching 40-50 μm in diameter) with a characteristic double-lens design (Fig. 3) [10]. The lenses are uniform in shape and appear as the scaled replicas of each other. The ratio between the lens thickness (t) and the lens diameter (L) estimated for a statistically significant set of micro-lenses was:

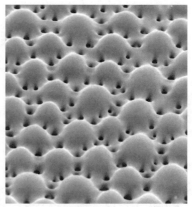

Figure 3. Array of microlenses on the surface of the dorsal arm plate in *O. wendtii.*

$$t/L = 0.9 \pm 0.05.$$

The degree of development of the lensar layer seems to correlate with the photosensitivity among several ophiocomid brittlestars that were studied. The layer is particularly pronounced in a highly photosensitive species, *Ophiocoma wendtii*, and it is absent in the light indifferent species, *Ophiocoma pumila* [10,17].

The optical properties of the microlens array were tested, using photolithography [10]. In these experiments, a film of a photosensitive material (positive photoresist) was illuminated through the isolated lensar layer (Scheme I). The illumination dose (I) was fixed slightly below the sensitivity level of the photoresist (I_0). As a result, the developed photoresist film was only affected precisely where the illumination dose increased due to the focusing activity of the lenses. The photoresist was placed at different distances h from the array. The photoresist films appeared to be selectively exposed under each micro-lens. Thus, the lensing effect was confirmed and mapped, and its quantitative characteristics (such as the position of the focal plane, d; the coefficient of the intensity enhancement by the micro-lenses, E; the angular selectivity, ϕ; the size of the focused beam at the focal plane, a_0; etc.) were experimentally determined from the values of h, L and the sizes of the spots in the photoresist, a, using basic equations of the geometric optics for a thick lens [22].

Scheme I.

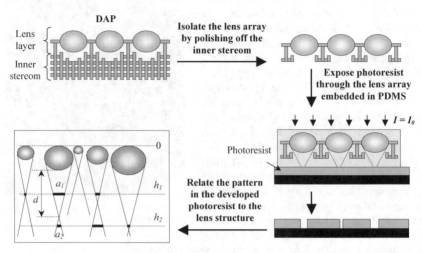

Our analysis showed that the average position of the focal plane is located at the distance d = 4–7 μm below the micro-lenses. The size of the spot in the focal plane a_0 is approximately 3 μm. The intensity of the incoming light is enhanced at the focal point by a factor of $E \sim$ 50. The angular selectivity ϕ is about 10°. These results presented the first experimental evidence that the micro-lens arrays are effective optical elements capable of significantly enhancing and guiding the light inside the tissue [10]. Calcitic micro-lenses of analogous structure were also reported in trilobite eyes [23-25].

3. Advantages of the Natural Micro-Lens Design

The optical characteristics of the external DAP layer obtained in the imaging experiments showed that brittlestar micro-lenses are in many aspects superior to their man-made analogs. This bio-optical system presents a spectacular example of Nature's ability to evolve sophisticated solutions even to complex technological problems.

We have identified a number of unique, advantageous properties and design strategies found in the brittlestar microlenses, which could be emulated in synthetic micro-lens arrays:

    a)   High fill factor lens array with characteristic micron-scale porous structure
    b)   Individually addressed crystalline lenses
    c)   Wide range transmission adjustment using pigment rearrangement
    d)   Characteristic lens geometry that minimizes spherical aberration
    e)   Fracture toughness due to the formation of an inorganic-organic composite
    f)   Crystallographic orientation along the optical axis of the lens crystal to eliminate birefringence
    g)   Angular selectivity
    h)   Cross-talk suppression by pigment distribution in pores

i)   Potential wavelength selectivity due to the pigment involvement
j)   Potential refractive index modification by gradient of specialized intracrystalline biomolecules

The importance of these properties in the improvement of the optical performance of micro-lenses is detailed below.

3.1. SIGNAL RECEPTION

It has been shown that the remarkable focusing properties of the bio-micro-lenses are used by the brittlestars in their natural habitat for survival purposes [18]. In a series of transmission electron microscopy (TEM) studies, Hendler and Byrne have identified neural bundles under the lenses, and suggested these to act as primary photoreceptors in the light signal detection [17]. This hypothesis was further confirmed in our lithographic experiments [10]. The nerve bundles are positioned in the pores of the stereom at the distance of about 5 μm from the lens, which corresponds well to the location of the experimentally identified focal point d. Moreover, the diameter of the nerve fibers correlates with the size a_0 of the focused beam in the focal plane.

Therefore, the arrays of lenses appear to form a sophisticated optical device, in which each component (a single micro-lens) is individually addressed by the detector positioned at the focal point of the lens. Due to the observed angular selectivity of these lenses and their different orientation, such device could act as a compound eye [26] that provides a wide field of view due to the non-planar arrangement of the lenses, which can be tuned on demand when the arm moves. The ways in which the signals detected by each receptor are processed and integrated remains unknown, and worthy of further investigation.

3.2. TRANSMISSION TUNABILITY

One of the most noticeable reactions to light in certain species of ophiocomid brittlestars is diurnal color change (Fig. 1) [18]. The change from a dark color during the day to a light color at night cannot be explained by camouflage, as has been shown experimentally [18]. Hendler and Byrne have shown that the channels in stereom contain chromatophore cells filled with pigment granules [17]. They proved that the diurnal color change is caused by the migration of these cells. The light microscopic sections of DAPs prepared from night and day samples showed that during the day, pigment-filled processes of the chromatophore beneath the epidermis cover the lenses. During the night the chromatophores withdraw into the stereom channels surrounding the lenses, deeper within the DAP. As a result, the intensity of light reaching the receptors is regulated by chromatophores depending on the illumination conditions; a function performed by an iris in a human eye. Similar process is utilized in so-called "transition" sunglasses. The behavior of *O. wendtii* suggests that the transmitted light signal is tuned to match the sensitivity level of the neural bundle and to optimize the reception [18].

The intensity-adjustment function of the chromatophore cells as well as the involvement of the sub-lens nerve bundles in photoreception (as opposed to a "diffuse" dermal reception), were corroborated by neurophysiological [27-28] data. Cobb and Hendler studied the

156

reaction of *O. wendtii* to light stimuli by direct recording from the nerve cord in brittlestar arms [29]. Nervous response of light-adapted, dark brown pigmented brittlestars has been monitored after the chemical disruption (by bleaching) of successive layers of the tissue. The removal of epidermal layer only (bleaching for less than 10 s) did not result in any change in the response compared to untreated, control specimens. Thus, "diffuse" photoreception by epithelial cells was ruled out. The removal of the pigment layer (bleaching for ~30 s) caused an approximately ten-fold increase in the response to light "ON", confirming the involvement of pigment in photoreception and its "sunglasses" function. The disruption of the sub-lens layer (bleaching for >45 s) resulted in no photic responses, consistent with the suggestion that the sub-lens neural bundles are the primary photoreceptors (Fig. 4).

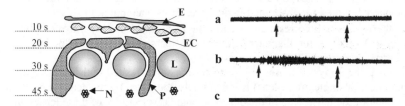

Figure 4. Left: Schematic presentation of the structure of the epidermis and dermis in the DAP. Lenses (L), pigment (P), cuticle of the epidermis (E), epithelial cells (EC) and nerve bundles (N) are indicated. Right: Recordings of the nervous response (right). Single arrows represent "Light 'ON'". Double arrows represent "Light 'OFF'". **a)** Response to light in a control, untreated specimen or specimen treated for <10 s. **b)** Response to light when the arm was bleached for 30 s. **c)** Response to light when the arm was bleached for >45 s. Adapted from [27].

In technical terms, the brittlestar micro-lenses is a tunable optical device that exhibits a wide-range transmission tunability achieved by controlled transport of radiation-absorbing intracellular particles. Other functions of the chromatocyte pigment may include the diaphragm action, and therefore, numerical aperture tunability; wavelength selectivity; minimization of the "cross-talk" between the lenses, and therefore, the improved angular selectivity.

3.3. ABERRATIONS

In an ideal lens, all rays of light would converge to the same point in the focal plane, forming a clear image (Fig. 5). The influences that cause different rays to focus to different points in real optical systems are called aberrations [22]. For a most common lens type, i.e. for lenses made with spherical surfaces, rays that are parallel to the optic axis but at different distances from the lens center do not converge on the same point. Rays that strike the outer edges of the lens are focused closer to the lens than rays that strike the inner portions of the lens (Fig. 5). This effect, called spherical aberration, results in a considerable image blurring and presents a serious technological problem in lens fabrication.

For a thick calcitic lens that has the size of the brittlestar micro-lenses and is formed by two spherical surfaces, the expected value of the light enhancement factor at the distance d from the lens would be $E_0 = 3\text{-}4$. The experimental value of E determined in the lithographic experiments was 15 times higher than the latter [10]. Therefore, the brittlestar micro-lenses must be significantly compensated for spherical aberration. Indeed, a close examination of the design of the brittlestar micro-lenses showed that they have a very peculiar shape. Only the top surface is spherical. The bottom surface has a characteristic aspherical form. The calculated shape of the ideal calcitic lens that is totally compensated for spherical aberration appeared to completely coincide with the shape of the bio-lens (Fig. 6) [10]. Moreover, the presence of pigment-filled chromatophores in the pores around the lenses would presumably block the light striking the outer portion of the lens, thus improving the operational size of the lens (L_0).

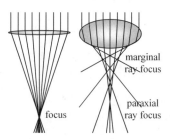

Figure 5. Ray tracing in an ideal, thin lens (left) and in a spherical, thick lens (right) that illustrates the concept of spherical aberration.

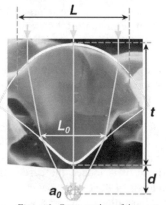

Figure 6. Cross-section of the brittlestar lens, showing its doublet structure delineated by one spherical and one aspherical surface. The bio-lens is superimposed with the calculated profile of an ideal lens that is compensated for spherical aberration (white contour). The match is striking.

Technological ways to minimize spherical aberration include: (i) the use of lens doublets instead of the single lens; (ii) bending one lens of the doublet into its best, aspherical (usually parabolic) form; (iii) the use of screens that disable the most problematic, exterior portions of the lens; and (iv) the use of high-index materials. Brittlestars seem to employ all of the above four approaches in their lens design. As a result, a remarkable level of

158

compensation for spherical aberration is achieved and significantly enhanced light converges into one point where the photoreceptor is positioned.

3.4. BIREFRINGENCE

Birefringence is the division of light into two components, which is found in materials that have two different indices of refraction in different crystallographic directions [22]. It is observed in crystals that have a crystallographic axis of higher symmetry, so-called uniaxial crystals. The direction of the axis of higher symmetry is optically unique, in that the propagation of light in this direction is independent of its polarization. This direction is called the optic axis of the material. Light propagating in any other direction will split into two beams that travel at different speeds.

Calcite is the classical example of a doubly refracting material. Its birefringence is extremely large, with indices of refraction of 1.66 and 1.48 along the optic axis and in the perpendicular direction, respectively. The use of calcite in the construction of optical lenses is, therefore, disadvantageous, as it will result in the formation of two images, unless the crystal is precisely oriented in the direction of the optical axis (Fig. 7a). Amazingly, in the brittlestar dorsal arm plate, it is indeed the case – the optical axis of the constituent calcite is oriented parallel to the lens axis and perpendicular to the plate surface (Fig. 7b). Since only the light striking the array perpendicular to the plate surface is effectively detected (rays striking the surface at an angle are stopped by pigment), the negative effect of birefringence is corrected and the receptor receives one clear signal.

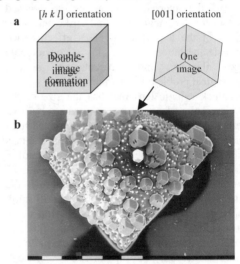

Figure 7. **a)** Schematic illustration of birefringence in calcite crystals. Double image is formed when the crystal is oriented in the general, {*h k l*} crystallographic direction. One image is formed when the crystal is oriented in the [001] direction, i.e. along the optical axis. **b)** Decoration of the DAP with synthetic calcite crystals that grow epitaxially on the surface. The experiment visualizes the crystallographic orientation of the biogenic substrate. The overgrown calcite crystals and therefore the lenses are oriented in the optic axis direction.

3.5. MECHANICAL PROPERTIES

Calcite is typically used in echinoderms for structural purposes [6,8,9]. This material is, however, highly brittle: cracks propagate easily along the {104} cleavage planes (Fig. 8, left). Organisms have evolved several sophisticated means to reinforce this intrinsically brittle material for skeleton construction. To that end, living organisms commonly employ organic-inorganic composite structures [8,9]. Addadi and Weiner suggested that specialized, intracrystalline macromolecules found inside biogenic calcite crystals of different origin are involved in the control of their mechanical properties [6,8,30,31]. These macromolecules are often adsorbed onto selective crystallographic planes in the crystal. It has been shown that in echinoderm calcite, the intracrystalline macromolecules interact specifically with crystallographic planes approximately parallel to the optic axis [6,8,30-32]. Positioned oblique to the cleavage planes (the optic axis forms an angle of about 45° with the cleavage planes), these macromolecules provide an effective crack-stopping mechanism by absorbing and deflecting the advancing cracks (Fig. 8, right). Such reverse fiber-reinforced composite exhibits reduced brittleness, increased plasticity, and fracture toughness [6,31].

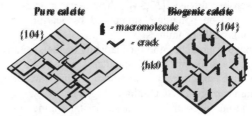

Figure 8. Schematic presentation of the crack-arresting function of specific intracrystalline macromolecules in biogenic calcites. See text for details.

As a result, the ossicles and associated micro-lenses formed by brittlestars are mechanically strong. It is also conceivable that intracrystalline macromolecules are involved in the regulation of the refractive index of calcite, thus further reducing the aberrations in the lens.

In summary, photosensitive brittlestars are impressively armed for light sensing. They form an array of nearly perfect optical lenses that are micron scale, lightweight, mechanically strong, aberration-free, birefringence-free, individually-addressed; that show a unique focusing effect, signal enhancement, intensity adjustment, angular selectivity, and photochromic activity. Together with neural receptors and intraskeletal chromatophores, these micro-lenses form an effective optical device that may function as a compound eye. Given appropriate neural integration, each DAP would have an effective visual field of ~10° surveying a different part of the object space [26]. Since *O. wendtii* has a large number of differently oriented DAPs, it could potentially extract a considerable amount of visual information about its environment.

4. Bio-Inspired Engineering

For the purpose of advancing the state-of-the-art optical technology striving towards the construction of a new generation of open space devices with variable field of view, tunable transmission, wavelength selectivity, etc., the brittlestar micro-lens arrays represent an inspirational example. We believe that the above biological principles, if understood correctly and creatively applied in technology, could well revolutionize our ability to make tunable, lightweight, porous micro-lens arrays for a wide variety of applications.

The following fabrication strategies were inspired by the principles involved in the formation of echinoderm calcitic structures.

4.1. *IN VITRO* NANOTECHNOLOGY EXPERIMENTS INVOLVING AMORPHOUS CALCIUM CARBONATE

A fascinating feature of echinoderm skeleton is that it is a multifunctional material composed of large single crystals with finely-tuned shape, micro-ornamentation, crystallographic orientation, and composition [8,9,20,21]. Single crystals with controlled micro-pattern are widely used in technology as components of various electronic, optical and sensory devices [33,34]. Their fabrication is, however, a complex, multistep process that could be potentially improved by learning from nature and introducing biological crystal growth techniques.

We have approached the mechanisms of the shape regulation in echinoderm calcite in the study of sea urchin larval spicules [35,36]. We have shown that the formation of final crystalline structures occurs through the transformation from the transient amorphous calcium carbonate (ACC) phase (Fig. 9).

For the purposes of bio-inspired crystal engineering, we identified two major elements of the larval spicule formation and used them in our synthetic effort:

(i) Amorphous calcium carbonate stabilized by specialized macromolecules is deposited in a preformed space and adopts its shape;

(ii) Oriented nucleation then occurs at a well-defined, chemically modified intracellular site and the crystallization front propagates through the amorphous phase, resulting in the formation of a single crystal with controlled orientation and predetermined microstructure.

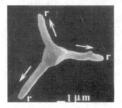

Figure 9. Sea urchin *Paracentrotus lividus* larval spicule (25 h embryo). Triradiate spicule is first deposited in an ACC form within a membrane-delineated compartment, inside a syncitium formed by specialized mesodermal cells. Within 20 h, an oriented calcite with the crystallographic *a*-axes parallel to the three radii nucleates (see a rhombohedral-shaped crystal at the center). The subsequent amorphous-to-crystalline transition results in the formation of a single crystal of the predetermined triradiate shape and crystallographic orientation [35].

4.1.1. *Oriented Nucleation of Calcite Controlled by Organically Modified Surfaces*
The crystallographically oriented growth of calcite is common in biological environments [6,8,20,35,37]. It is generally believed that crystal nucleation is templated by membranes in the form of vesicles, syncitia, and cells, which are usually "primed" with specialized, acidic macromolecules (rich in Asp, Glu, Ser, Thr, often sulfonated or phophorylated) [8,30,32,37]. The control of crystallographic orientation is achieved by stereochemical recognition at the organic-inorganic interface, conceivably by virtue of a match between the structures of the organic surface and that of a particular crystal plane [3,6,38].

In our effort to mimic the biological process of oriented crystallization, we templated calcite nucleation by constructing specially-designed organic surfaces. Self-assembled monolayers (SAMs) of alkanethiols terminated in functional groups that have a direct biological relevance (CO_2^-, SO_3^-, PO_3^{2-}, OH) were used as substrates for calcium carbonate deposition [38,39]. We showed that such templates induced a remarkable level of control of crystallographic orientation of calcite crystals. The crystallographic orientations of the crystals were distinct and homogeneous for each surface, but different on different surfaces. An even higher level of control was achieved in the experiments involving micropatterned SAMs [40]. By adjusting the geometry and sizes of the features in the SAM pattern, the concentration of the solution, and the functionality of the surface of the SAM, we could fabricate arrays of oriented crystals with the controlled density of nucleation in defined locations (Fig. 10) [7].

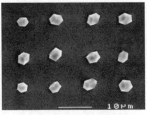

Figure 10. Oriented nucleation of an array of calcite crystals templated by a patterned organic substrate. Islands of $HS(CH_2)_{15}CO_2H$ SAM supported on Ag were used. Crystals grow uniformly in the [012] direction.

4.1.2. *Stabilization of Amorphous Calcium Carbonate*
Amorphous calcium carbonate (ACC) is highly unstable. If not stabilized, this phase rapidly transforms into crystalline polymorphs [41-44]. The formation of ACC is, however, common in biological systems [35,36,45,46]. We have shown that biogenic ACC is stabilized by means of specialized macromolecules rich in hydroxyamino acids, glycine, glutamate, phosphate and polysaccharides [45, 46]. These macromolecules extracted from the biological ACC tissue and introduced as an additive into saturated calcium carbonate solution induced the formation of stabilized ACC *in vitro*.

The stabilization of ACC can also be achieved using surface-mediated processes. Our results show that significantly disordered organic substrates bearing phosphate and hydroxyl groups, suppress the nucleation of calcite and induce the formation of a metastable ACC layer from highly saturated solutions [47]. Such disordered organic surfaces can be generated using a mixture of alkanethiols of different lengths, functionalized with PO_3H, OH and CH_3 groups and supported on metal substrates (we will refer to this surface as to "ACC-inducing"). The difference in the length of the carbon chain in the thiol molecule

162

prevents the formation of an ordered interfacial layer, and therefore, the formation of ordered nucleation sites. The deposited ACC layer is stabilized for 1-2 hrs and then transforms into a polycrystalline film.

When the substrate is engineered in such a way that ordered SAMs that induce oriented calcite nucleation are patterned *into* the ACC-inducing background, the formation of the ACC layer is followed by the controlled oriented nucleation at the defined locations of the SAM and the transformation of the patterned ACC layer into an array of oriented calcite crystals (Fig. 11).

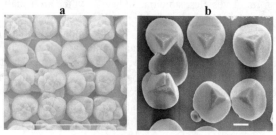

Figure 11. Surface-induced formation of ACC (**a**) and its transformation into an array of oriented calcite crystals (**b**). Scale bars are 100 μm and 10 μm, respectively. An ACC-inducing surface was patterned with a square array of 1 μm dots of HS(CH₂)₁₁PO₃H₂ serving as nucleation sites. The *c*-axes of the crystals form a constant angle of ~20° with the surface normal, as expected for the oriented growth of calcite on a PO₃-terminated SAM.

4.1.3. *Fabrication of Large Single Calcite Crystals with Controlled Micropattern and Orientation*

The above studies show that ACC may be successfully used as the transient phase for calcite crystallization in an artificial system. Moreover, if nucleation sites are imprinted onto the ACC-inducing surface, the ACC transforms into oriented crystals whose orientation is controlled by the nucleating region of the template.

In our new approach to the fabrication of micropatterned single crystals, we apply the biomineralization principles summarized in Chapter 4.1; that is, the use of the controlled amorphous-to-crystalline transition initiated at a well-defined nucleation site and taking place within an ornate reaction volume that determines the shape of the mineral [47]. Figure 12 outlines the experimental procedure. An arbitrary photoresist micropattern (usually an array of isolated posts with feature sizes <10 μm) was formed on a glass slide using routine photolithographic procedures. The micropatterned substrate was primed with a disordered phosphate-, methyl- and hydroxy-terminated monolayer that induces the formation of amorphous $CaCO_3$. One nanoregion of the SAM of $HS(CH_2)_nA$ (A = OH, CO_2H) serving as calcite nucleation site was integrated into each template (Fig. 12a). When placed in a supersaturated solution of calcium carbonate, these organically modified templates induced the deposition of a micropatterned ACC mesh in the interstices of the framework. Oriented nucleation of calcite then occurred at the SAM nanoregion, followed by the propagation of the crystallization front through the ACC film (Fig. 12b).

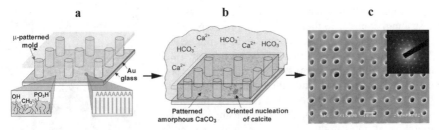

Figure 12. **a)-b)** Schematic presentation of the fabrication procedure of micropatterned crystals (see text for details). **c)** Magnified scanning electron micrograph of the resulting crystal, showing its microstructure. Inset: large-area transmission electron diffraction, confirming that the section is a single crystal of calcite oriented along the optic axis.

This new bio-inspired crystal engineering strategy made it possible, for the first time, to directly fabricate millimeter-size single crystals with a predetermined sub-10-micron pattern and controlled crystallographic orientation (Fig. 12c). We showed that in addition to their "shaping" function and the control of the ACC formation and nucleation, organically modified 3D templates act as the stress release sites as well as discharge sumps for excess water and impurities during crystallization [47]. The described mechanism of the amorphous-to-crystalline transition may have direct biological relevance and important generic implications in the fabrication of defect-free, micropatterned crystalline materials for a wide variety of applications.

The design of these synthetic calcite crystals is remarkably similar to the perforated stereom structure of echinoderm skeletal elements – their biological analogs. We are currently exploring the ways to synthesize single crystalline micro-lens arrays using the above strategy. Towards this goal, we are creating the micropatterned frameworks with the top surfaces bearing arrays of concave structures that will define the shape and the curvature of the lenses.

4.2. FABRICATION OF TUNABLE POROUS MICRO-LENS ARRAYS USING INTERFERENCE LITHOGRAPHY

An interesting feature of the brittlestar micro-lens arrays is the presence of pores surrounding the lenses. We discussed in Chapter 3.2 that these pores are important functional elements of the biological optical device, as they are used for the transport of pigment that regulates the illumination dose reaching the lens.

We have developed a novel, simple approach that uses multi-beam interference lithography to create porous hexagonal micro-lens (1-8 μm in diameter) arrays from photoresist materials [48]. In our experiments, we used a continuous wave diode-pumped solid state laser to photopolymerize a negative tone resist SU8. The optics setup was similar to that described by Turberfield *et al.* [49]. The physical basis for the process is detailed below. When the interference light is transferred into a negative tone photoresists during exposure, a periodic pattern of strongly and weakly exposed regions is generated. The highly exposed regions are then polymerized and the unexposed regions are dissolved away to reveal the holes. When the difference between the adjacent strongly and weakly exposed regions is close to a certain threshold value, the gradual change in the intensity between the regions produces the lens-like topography in the photoresist film combined with holes.

The appearance of thus synthesized micro-lens arrays (Fig. 13) is strikingly similar to their biological prototype shown in Fig. 3. The lens size, shape, symmetry and connectivity are controlled by beam wave vectors and their polarizations; while the pore size is adjusted by laser intensity, exposure time and the concentration of a photosynthesizer in the resist. We showed that these synthetic micro-lenses are capable of focusing light. The incorporation of holes in the lens array provides means for the transport of photoradiation-absorbing liquids and, therefore, for transmission and numerical aperture tunability and wavelength selectivity.

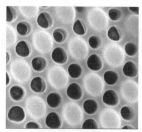

Figure 13. Micro-lens array with integrated pores fabricated by multi-beam lithography.

5. Conclusion

Multidisciplinary groups involving materials scientists, chemists, physicists, biologists work together trying to understand the mechanisms controlling the formation of elaborate structures of biological minerals. We believe that further studies of biological systems will increase our understanding of how organisms evolved their sophisticated optical structures for survival and adaptation and will provide additional materials concepts and design solutions. Ultimately, these biological principles will improve our current capabilities to fabricate optical elements and contribute to the construction of novel, adaptive, micro-scale optical devices.

We thank S. Weiner, L. Addadi, M. Byrne, J. Cobb, G. Lambert, A. Tkachenko, G. M. Whitesides, S. Yang, D. R. Hamann, M. Megens for their contribution to this work.

References

1. Alper, M., Calvert, P.D., Frankel, R., Rieke, P.C. and Tirrell, D.A. (1991) *Materials synthesis based on biological processes,*. Materials Research Society, Pittsburgh.
2. Braun, P.V., Osenar, P. and Stupp, S.I. (1996) Semiconducting superlattices templated by molecular assemblies, *Nature* **380**, 325-328.
3. Mann, S., Archibald, D.D., Didymus, J.M., Douglas, T., Heywood, B.R., Meldrum, F.C. and Reeves, N.J. (1993) Crystallization at inorganic-organic interfaces - Biominerals and biomimetic synthesis, *Science* **261**, 1286-1292.
4. Heuer, A.H., Fink, D.J., Laraia, V.J., Arias, J.L., Calvert, P.D., Kendall, K., Messing, G.L., Blackwell, J., Rieke, P.C., Thompson, D.H., Wheeler, A.P., Veis, A. and Caplan, A.I. (1992) Innovative materials processing strategies - a biomimetic approach, *Science* **255**, 1098-1105.
5. Mann, S. and Ozin, G.A. (1996) Synthesis of inorganic materials with complex form, *Nature* **382**, 313-318.
6. Addadi, L. and Weiner, S. (1992) Control and design principles in biological mineralization. *Angew. Chem.-Int. Edit. Engl.* **31**, 153-169.

7.   Aizenberg, J., Black, A.J. and Whitesides, G.M. (1999) Control of crystal nucleation by patterned self-assembled monolayers, *Nature* **398,** 495-498.

8.   Lowenstam, H. A. and Weiner, S. (1989) *On Biomineralization,* Oxford Univ. Press, Oxford.

9.   Wainwright, S.A., Biggs, W.D., Currey, J.D. and Gosline, J.M. (1976) *Mechanical design in organisms,* John Wiley and Sons, New York.

10.   Aizenberg, J., Tkachenko, A., Weiner, S., Addadi, L. and Hendler, G. (2001) Calcitic microlenses as part of the photoreceptor system in brittlestars, *Nature* **412,** 819-822.

11.   Cattaneo-Vietti, R., Bavestrello, G., Cerrano, C., Sara, M., Benatti, U., Giovine, M. and Gaino, E. (1996) Optical fibres in an Antarctic sponge, *Nature* **383,** 397-398.

12.   Sundar, V.C., Yablon, A.D., Grazul, J.L., Ilan, M. and Aizenberg, J. (2003) Fiber-optical features of a glass sponge, *Nature* **424,** 899-900.

13.   Sarikaya, M., Fong, H., Sunderland, N., Flinn, B.D., Mayer, G., Mescher, A. and Gaino, E. (2001) Biomimetic model of a sponge-spicular optical fiber - mechanical properties and structure, *J. Mater. Res.* **16,** 1420-1428.

14.   Hyman, L.H. (1955) *The invertebrates: Vol. 4, Echinodermata,* McGraw-Hill, New York.

15.   Yoshida, M., Takasu, N. and Tamotsu, S. (1984) Photoreception in echinoderms, in M.A. Ali (eds.), *Photoreception and vision in invertebrates,* Plenum, New York, pp. 743-771.

16.   Millot, N. (1975) The photosensitivity of echinoids, *Adv. Mar. Biol.* **13,** 1-52.

17.   Hendler, G. and Byrne, M. (1987) Fine structure of the dorsal arm plate of *Ophiocoma wendti:* Evidence for a photoreceptor system (Echinodermata, Ophiuroidea), *Zoomorphology* **107,** 261-272.

18.   Hendler, G. (1984) Brittlestar color-change and phototaxis (Echinodermata: Ophiuroidea: Ophiocomidae), *PSZNI Mar. Ecol.* **5,** 379-401.

19.   Cowles, R.P. (1910) Stimuli produced by light and by contact with solid walls as factors in the behavior of ophiuroids, *J. Exp. Zool.* **9,** 387-416.

20.   Donnay, G. and Pawson, D.L. (1969) X-ray diffraction studies of echinoderm plates, *Science* **166,** 1147-1150.

21.   Ameye, L., Hermann, R., Wilt, F. and Dubois, P. (1999) Ultrastructural localization of proteins involved in sea urchin biomineralization, *J. Histochem. Cytochem.* **47,** 1189-1200.

22.   Flint, H.T. (1936) *Geometrical optics,* Methuen and Co, London.

23.   Clarkson, E.N.K. and Levi-Setti, R. (1975) Trilobite eyes and the optics of Des Cartes and Huygens, *Nature* **254,** 663-667.

24.   Gal, J., Horvath, G., Clarkson, E.N.K. and Haiman, O. (2000) Image formation by bifocal lenses in a trilobite eye?, *Vision Res.* **40,** 843-853.

25.   Towe, K.M. (1973) Trilobite eyes: Calcified lenses in vivo, *Science* **179,** 1007-1010.

26.   Land, M.F. (1981) Optics and vision in invertebrates, in H. Autrum (eds.), *Comparative physiology and evolution in invertebrates B: Invertebrate visual centers and behavior I*, Springer, Berlin, pp. 471-592.

27.   Cobb, J.L.S. and Hendler, G. (1990) Neurophysiological characterization of the photoreceptor system in a brittlestar, Ophiocoma wendtii (Echinodermata: Ophiuroidea), *Comp. Biochem. Physiol.* **97A,** 329-333.

28.   Stubbs, T.R. (1982) The neurophysiology of photosensitivity in ophiuroids, in J.M. Lawrence (eds.), *Echinoderms: Proceedings of the International Conference, Tampa Bay,* Balkema, Rotterdam, pp. 403-408.

29.   Johnsen, S. (1997) Identification and localization of a possible rhodopsin in the echinoderms *Asterias forbesi* (Asteroidea) and *Ophioderma brevispinum* (Ophiuroidea), *Biol. Bull.* **193,** 97-105.

30.   Berman, A., Addadi, L., Kvick, Å., Leiserowitz, L., Nelson, M. & Weiner, S. (1990) Intercalation of sea urchin proteins in calcite: Study of a crystalline composite material, *Science* **250,** 664-667.

31.   Addadi, L., Aizenberg, J., Albeck, S., Berman, A., Leiserowitz, L. & Weiner, S. (1994) Controlled occlusion of proteins - a tool for modulating the properties of skeletal elements, *Mol. Cryst. Liq. Cryst. Sci. Technol. Sect. A-Mol. Cryst. Liq. Cryst.* **248,** 185-198.

32.   Albeck, S., Aizenberg, J., Addadi, L. & Weiner, S. (1993) Interactions of various skeletal intracrystalline components with calcite crystals, *J. Am. Chem. Soc.* **115,** 11691-11697.

33.   Gonis, A. (2000) *Nucleation and Growth Processes in Materials*, Materials Research Society, Boston.

34.   Vere, A.W. (1988) *Crystal Growth: Principles and Progress*, Plenum, New York.

35.   Beniash, E., Aizenberg, J., Addadi, L. and Weiner, S. (1997) Amorphous calcium carbonate transforms into calcite during sea urchin larval spicule growth, *Proc. R. Soc. Lond. Ser. B-Biol. Sci.* **264,** 461-465.

166

36. Beniash, E., Addadi, L. and Weiner, S. (1999) Cellular control over spicule formation in sea urchin embryos: A structural approach, *J. Struct. Biol.* **125**, 50-62.
37. Aizenberg, J., Hanson, J., Koetzle, T.F., Leiserowitz, L., Weiner, S. & Addadi, L. (1995) Biologically induced reduction in symmetry - a study of crystal texture of calcitic sponge spicules, *Chem.-Eur. J.* **1**, 414-422.
38. Aizenberg, J., Black, A.J. and Whitesides, G.M. (1999) Oriented growth of calcite controlled by self-assembled monolayers of functionalized alkanethiols supported on gold and silver, *J. Am. Chem. Soc.* **121**, 4500-4509.
39. Han, Y.-J. and Aizenberg, J. (2003) Face-selective nucleation of calcite on self-assembled monolayers of alkanethiols: Effect of the parity of the alkyl chain, *Angew. Chem. Int. Ed.* **42**, 3668-3670.
40. Xia, Y.N. and Whitesides, G.M. (1998) Soft lithography, *Annu. Rev. Mater. Sci.* **28**, 153-184.
41. Koga, N., Nakagoe, Y.Z. and Tanaka, H. (1998) Crystallization of amorphous calcium carbonate, *Thermochim. Acta* **318**, 239-244.
42. Gower, L.B. and Odom, D.J. (2000) Deposition of calcium carbonate films by a polymer-induced liquid-precursor (PILP) process, *J. Cryst. Growth* **210**, 719-734.
43. Raz, S., Weiner, S. and Addadi, L. (2000) Formation of high-magnesian calcites via an amorphous precursor phase: Possible biological implications, *Adv. Mater.* **12**, 38-41.
44. Sawada, K. (1997) The mechanisms of crystallization and transformation of calcium carbonates, *Pure Appl. Chem.* **69**, 921-928.
45. Aizenberg, J., Lambert, G., Addadi, L. and Weiner, S. (1996) Stabilization of amorphous calcium carbonate by specialized macromolecules in biological and synthetic precipitates, *Adv. Mater.* **8**, 222-225.
46. Aizenberg, J., Lambert, G., Weiner, S. and Addadi, L. (2002) Factors involved in the formation of amorphous and crystalline calcium carbonate: A study of an ascidian skeleton, *J. Am. Chem. Soc.* **124**, 32-39.
47. Aizenberg, J., Muller, D.A., Grazul, J.L. and Hamann, D.R. (2003) Direct fabrication of large micropatterned single crystals, *Science* **299**, 1205-1208.
48. Yang, S., Megens, M. and Aizenberg, J. (2003) Fabrication of biomimetic microlens arrays with integrated pores by interference lithography, *Unpublished data.*
49. Campbell, M., Sharp, D.N., Harrison, M.T., Denning, R.G. and Turberfield, A.J. (2000) *Nature* **404**, 53-55.

4. Tissue Engineering of Mineralized Tissues

INKJET PRINTING FOR BIOMIMETIC AND BIOMEDICAL MATERIALS

PAUL CALVERT[*], YUKA YOSHIOKA and GHASSAN JABBOUR

University of Arizona, Tucson AZ 85721
[*]*Now at Dept. of Textile Sciences, University of Massachusetts at Dartmouth, N. Dartmouth, MA 02747*

Abstract

Growth of biological tissues often occurs by a layer of cells building new tissue layer-by-layer. Each new layer is the result of the reaction between a series of reagents and catalysts expressed by the cell layer. The nozzles of inkjet printheads have dimensions similar to those of cells and so it is possible to imagine a versatile inkjet printer also building tissues by depositing a series of drops of various reagents. This paper describes preliminary tests of this approach, the use of inkjet printing to form lines and layers of polymers, reaction between successive drops to produce layers of gels about 100nm thick, addition of enzymes into these gels and there use to produce mineralized structures.

1. Introduction

Inkjet printing is familiar as a method for printing on paper. In recent years there has been much interest in applying it to the deposition of materials for a variety of applications, including metallic and polymer conductors, organic semiconductors, ceramics, DNA arrays, waxes and polymers [1]. The advantages of this method, compared to other printing methods, are that it is non-contact, can be applied to almost any low-viscosity liquid and can form very thin films or build thick layers. Limitations include a lateral resolution of 10 microns or more, formation of rough films from joined dots, difficulties in keeping the nozzles clear and non-uniformities due to drying processes.

In this paper we discuss the use of inkjet printing as an analog for extracellular formation of biological materials. The concept is to build tissue-like materials by the sequential printing of reactive and catalytic solutes. Reaction between these will allow complex combinations

R.L. Reis and S. Weiner (eds.),
Learning from Nature How to Design New Implantable Biomaterials, 169-180.

of materials to be formed layer by layer with the incorporation of structural or active elements as desired.

2. Inkjet Printing as a Biological Model

Many biological growth processes occur as an extra-cellular formation of an insoluble solid from dissolved components. Where a monolithic solid structure is being formed or expanded, as in the growth of bone, tooth or tendon, the process is performed by a layer of cells that deposit successive sheets of solid and then move back. In many cases, such as bone [2,3], the solid does not form right against the cell membrane but there is a gap of a few microns with unstructured matrix across which solutes must be transported prior to crystallization. In the case of enamel, [4], growing crystal do seem to be in local contact with the cell membrane but even so, much of the densification occurs several microns behind the growth front. The presence of an unstructured or liquid layer ahead of a growth front would seem to be necessary to avoid the formation of a dense impermeable skin with trapped voids.

Given this, there must be a mechanism for cells to produce soluble products for which solidification is delayed until they have traveled by diffusion to some distance from the cell surface. Immediate precipitation would result in occlusion. Collagen for instance, is formed from procollagen. The soluble protein is exported by the cell and is then cleaved by an extracellular enzyme to form tropocollagen triple helices, which then assemble into collagen fibrils. While the details are unknown, at least to us, there is clear potential for control of the site of collagen formation by controlling the concentrations and the relative diffusion and reaction rates for the components of this system.

Many biological tissues are formed extracellularly as a result of serial production of reagents by a layer of cells that deposit a layer of new tissue and move back to repeat the process. These reagents, including proteins and minerals, pass through the cell membrane in soluble form and then interact externally to form gels, fibers and minerals. One model for this might be an array of nozzles at 20-50 micron spacing capable of expelling a programmed series of reagents into a layer of liquid.

Inkjet nozzles are typically 20-50 microns in diameter. An inkjet print head capable of delivering 4-8 reagents, in lieu of colors, and able to move rapidly over an area can be seen as performing a similar function to a layer of cells. In exploring the ability of sequential inkjet printing may we come to understand better the importance of time and space in tissue formation.

3. Aspects of Inkjet Printing

3.1. THE PRINTING PROCESS

Two main methods of drop formation are used in drop-on-demand printers. Many systems depend on a piezoelectric actuator, which produces a pressure pulse in a column of ink and ejects a drop. Other printers use a small heating element to form a vapor bubble in the ink, which expels a drop as it expands.

Most printheads intended for commercial use have piezoelectric drivers but we have found that the thermal system is relatively rugged and simple and so is convenient for use in the laboratory [1]. Figure 1 shows our system with a thermal printhead attached to a pulse generator, pressure regulator and xyz stage.

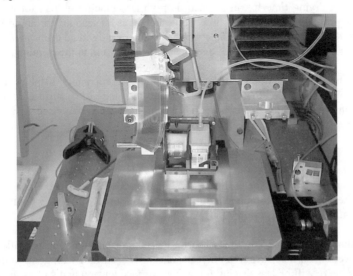

Figure 1. A thermal inkjet printhead connected to drive electronics to fire a single nozzle and mounted over an xyz stage.

Droplet size and uniformity have been a major concern to manufacturers of inkjet print heads. The smaller the size, the higher the resolution available for printed images. Uniformity has become a special concern as inkjet printing becomes used in electronic displays. In addition the process often produces satellites, small droplets which form during the breakoff of the main droplet from the nozzle. A discussion of these questions has recently been published [5].

3.2. DROPLET DRYING AND COALESCENCE

In printing materials one wants to be able to print lines or layers onto a flat substrate. When inkjet printing onto transparencies it is normal to use pre-coated substrates with an absorbing polymer layer to prevent the ink coalescing into bigger drops. In making devices, we would like to be able to print onto uncoated glass or plastic and then to overprint further layers of material. For this we need some understanding of how droplets will interact on the surface. Other aspects of rheological effects on droplet delivery have recently been discussed [6].

A single drop on a surface will dry at a rate, which depends on the solvent evaporation rate. A sheet of water in still air at room temperature evaporates at a rate of about 1.1 microns/sec. The drying time for an isolated drop will be much less than for a film or an array of drops printed on a surface, which would be about 30 seconds for an array formed from droplets of 25 micron initial radius. Also the drying of individual drops in an array will depend on their position relative to the edge and on their position in the printing sequence.

Consider two droplets arriving on a surface so that they are in contact. If the contact line is not pinned on the surface, they will fuse under the action of surface tension. The approximate timescale for this process depends on surface tension and viscosity and is given by:

$$\tau \sim 2.r^2.\eta/\sigma.h \sim 10^{-5} \text{ sec} \qquad (1)$$

where r is the radius of the sessile drop, η is the liquid viscosity, σ is the surface tension and h is the drop height, usually comparable to r. Thus drop shape will usually equilibrate long before evaporation occurs.

Since the fusion time is generally much shorter than the drying time, drops will fuse on a clean, partly wetting surface such as a polyester sheet. On a surface with good wetting, such as clean glass, there will be less driving force from surface tension and so less tendency for drops to form into a single large drop and more to form a line or film.

As drying starts, the contact line is often pinned by surface roughness or by deposition of solute or particles at the contact line [7,8]. This can prevent fusion of adjacent droplets.

As a practical matter, we find that suspensions of particles can readily be raster printed as lines and areas on glass by warming the substrate to the point where the drying time is less than the time to print a single line. This allows individual droplets to fuse but prevents large liquid droplets from forming. On plastic substrates, there is more of a tendency for drops to flow together. For instance, adjacent lines will fuse into a single line. We have printed areas by depositing an array of separate drops, allowing this to dry and then filling in the gaps with a second array.

The "coffee ring" effect has become of particular concern in the printing of polymer and small-molecule inks for light-emitting displays. Inkjet printing lends itself to depositing the different colored pixels needed for a red-green-blue display. A "coffee ring" is a ring of thicker deposit around the edge of the drop. This process has been analyzed for larger drops and is attributable to a higher evaporation rate at the contact line of a pinned drop, compared to the center of the drop surface [9,10]. As a result solution flows to the edge of the drop and the excess solute or suspended particles become deposited at the periphery. Empirical studies have found methods to control this non-uniformity, apparently by speeding up the drying process, but the effect is not fully understood [11].

3.3. SMOOTHING PRINTED FILMS

A film printed from a succession of drops will be rough on a scale comparable to the film thickness, typically about 100nm. In a polymer, this roughness can be smoothed out by annealing at a temperature above the glass transition. Figure 2 shows a film of PEDOT/PSS conducting polymer, with 50% glycerol, printed and dried at 25 °C, showing a surface roughness determined by a Wyko profilometer as 81 nm. After annealing at 110 °C for 1 hour, the roughness has decreased to 3.9nm, comparable to that seen in spin-coated films, Figure 3. We believe that the glycerol-plasticized film smooths by the action of surface tension.

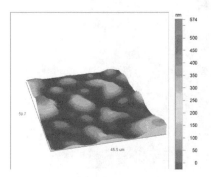

Figure 2. Topography of a printed film of conducting polymer as printed.

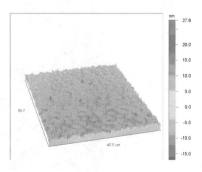

Figure 3. Film from figure 2 after annealing at 110°C.

3.4. PRINTING ON ABSORBING SURFACES

Polyester transparencies for inkjet printing are normally coated with an absorbing polymer "receptor" layer to control the spreading of the ink. In biological growth, extracellular matrix is often formed prior to formation of structural tissue, presumably to provide a degree of control that would not be available in a fluid layer. In mimicking this using an inkjet printer, one would want to deposit either a gel layer or a porous layer in order to provide a controlled reaction space and then add a subsequent layer to this once the structural material had formed. Figure 4 shows such a layer of alumina particles printed onto substrate. The alumina was dispersed in water using an anionic polymer (polystyrenesulfonate-co-maleic acid) and then was immobilized by overprinting with a cationic polymer, which binds the particles together and prevents redispersion during subsequent printing steps.

4. Printing of Multiple Materials

4.1. DROPLET REACTION

The particular interest of our work has been to explore the printing of multiple inks which react to form a product material. In one example of this we have successively printed a water-soluble amine (Jeffamine T403) and a water-soluble epoxy (ethyleneglycoldiglycidylether, glyceroldiglycidylether or polyethyleneglycoldiglycidylether), in order to form a water-swellable crosslinked layer. In the case of one droplet printed on top of another dried drop, the solvent would first redisperse the first ink and form a two-layer droplet.

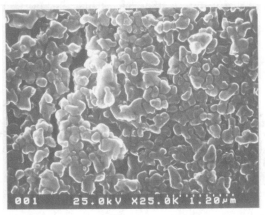

Figure 4. Layer of 0.18 micron alumina particles printed onto a polymer substrate as an adsorbing layer and gelled.

The two inks then can diffuse and react, unless reaction is very rapid and forms a barrier to further diffusion. For most epoxies, reaction times are minutes or hours and so diffusive mixing occurs first and a uniform gel is expected. Figure 5 shows an image of fluorescently-labelled serum albumin co-printed with the amine component and immobilized within such an epoxy gel. The gel and protein are not removed by washing.

Figure 5. Fluorescent serum albumin immobilized in an epoxy gel layer on silicon, resistant to washing.

Insoluble, attached gels can also be formed by self-assembly of complementary polyelectrolytes. Thus figure 6 shows a gel formed by sequentially printing a solution of anionic polymer (R478, a red polymeric dye) and cationic polymer (polydiallyldimethylammonium chloride). The final structure shows precipitated threads of polymer complex and must depend on the ratio of charged polymers, the molecular weights, the order in which the polymers are printed and the degree of wetting of the substrate.

When the anionic protein serum albumin is printed and the overprinted with cationic polymer, a uniform insoluble sheet is formed, figure 7.

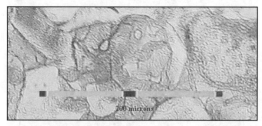

Figure 6. Gelled polymer layer by sequential printing of 2% solutions of cationic and anionic polymer onto a glass surface, after washing.

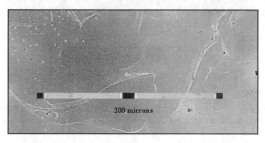

Figure 7. Gelled fluorescent serum albumin layer formed by sequentially printing the protein and cationic polymer solutions.

4.2. MINERALIZATION-ACTIVE GEL MATRICES

Using the epoxy gel system we have immobilized alkaline phosphatase in lines of gel 100 microns thick formed with an extrusion system [12]. These were then immersed in a solution containing calcium ions and a phosphate ester. Figure 8 shows strips of such gel in p-nitrophenylphosphate. Both the gel and the surrounding solution become yellow with released p-nitrophenol, showing that the enzyme is still active. Microscopy of these strips after immersion in phenylphosphate and calcium chloride shows that they are covered with a layer of mineral, figure 9, which appear to be plate-like hydroxyapatite. EDS analysis gives a Ca:P ratio of 2.5 on uncoated samples, the high value probably being due to adsorbed calcium chloride from the growth solution. The mineral can also be detected inside the gel. In single-phase gels from Jeffamine T403 and polyethyleneglycoldiglycidylether, the Ca:P ratio is 1:1 suggesting that brushite forms, whereas two-phase gels from Jeffamine T403 and glyceroldiglycidylether give a Ca:P ratio of 1.7:1 suggesting that hydroxyapatite is formed. In either case, the scale of the mineral is too fine to be resolved.

Figure 8. Epoxy gel strips without (left) and with immobilized alkaline phosphatase after immersion in p-nitrophenylphosphate. Both the gel and solution on the right become yellow due to the action of the enzyme to form p-nitrophenol.

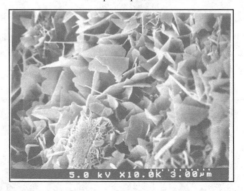

Figure 9. Plate-like crystals growing on a gel layer with immobilized alkaline phosphatase after immersion in phenylphosphate and calcium chloride. The interior of the gel also contained fine scale mineral.

The same gel and enzyme system was inkjet printed in lines about 1 micron thick on various substrates and were again immersed in calcium and phosphate ester. In this case mineral crystals did form on the substrates but not specifically on the lines, figure 10. We believe that these, much thinner, lines led to a shifted balance between formation and diffusion of phosphate ions such that there was no localized high concentration in the lines and no localized nucleation.

This system is a complex balance between diffusion of the phosphate ester into the line, conversion to phosphate and diffusion out. Nucleation and precipitation will occur, either if preferential nucleation sites are also present in the line or if the local concentration of phosphate is much higher than that in the surrounding solution. As a simple rule-of-thumb, if the characteristic time for diffusion out of the line is much less than the half-time for conversion of the ester to phosphate, no precipitation will occur. The diffusion time is $x^2/2D$, where x is the line thickness and D is the diffusion coefficient of phosphate in the gel. For a 1 micron thick line, this will be less than a second, whereas it will be hundreds of seconds for a line 100 microns thick. Thus we see that processes similar to extra-cellular tissue formation can be locally directed by localization of reagents and catalysts if diffusional processes are controlled.

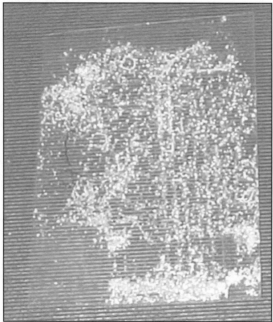

Figure 10. Mineralization on thin inkjet printed lines with immobilized alkaline phosphatase. Mineralization now occurs on all surfaces and is not localized to the lines.

4.3. MODELING OF ENZYMATIC PROCESSES IN THIN LAYERS

It is possible that the progressive formation of tissue with distance from the active cell layer is due to slow diffusion of reactive species. This could result, for instance, in a decreasing concentration of an inhibitor away from the cell surface. Given tissue growth rates of a few millimeters per day, control mechanisms should operate on the scale of about a few hundred seconds if the active zone is 10 microns wide. Diffusion coefficients for molecules in aqueous solution range from about 10^{-5} to 10^{-7} cm^2/sec as the molecular weight rises from 50 to 500,000. For a film of thickness 10 microns, these give relaxation times for equilibration of concentration of about 0.5 to 5 seconds. Thus a diffusion-based control system would require diffusion coefficients reduced by two orders of magnitude. This could be the case in a gel matrix, if either the diffusing species bound to the gel or were large compared to the pore size of the gel. Either of these may be true for proteins diffusing in extracellular matrix.

As an illustration of this model, figure 11 shows the formation of a product in a 100 micron layer of matrix against a surface (cell layer), which acts as a constant source. Also shown is the same system with a slowly diffusing inhibitor, which results in a reduced level product adjacent to the source. In order to produce a distribution similar to bone growth, with product formation at a distance of a few microns from the cell layer, the inhibitor can be produced slightly before the reagents so that a zone of no reaction is formed, as shown in figure 12. Thus, given a suitable combination of reaction and diffusion rates and release times, it is possible to control deposition in depth as seems to occur in tissue formation.

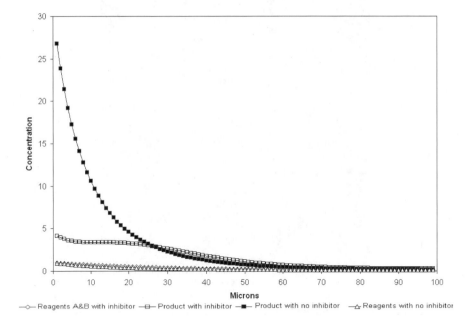

Figure 11. Product and reagent distribution in a 100 micron layer with co-diffusion for 6 seconds of reagents (D= 10^{-6}cm^2/sec) and a slowly diffusing inhibitor (D= 10^{-9}cm^2/sec).

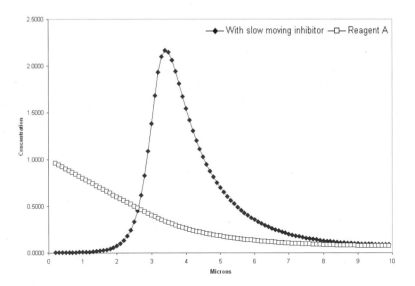

Figure 12. Product and reagent distribution in a 10 micron layer after 3 seconds of diffusion of an inhibitor (D= 10⁻⁹cm²/sec) followed by diffusion of two reagents (D= 10^{-6} cm²/sec), with rapid reaction A+B->C.

5. Conclusions

We have shown that inkjet printing can be used to deposit multiple layers of reacting materials and so could be used to mimic biological growth. The form of the product will depend on the interaction of the printed materials with the substrate and with each other. Effects of surface tension and of timing are important. The product will depend on a combination of diffusion and chemical reaction which may need to be modeled in order to achieve detailed understanding. This approach may also shed light on the role of time, space and diffusional processes on biological tissue formation.

References

1. Calvert, P. (2001) Inkjet printing for materials and devices, *Chem. Mater.* **13**, 3299-3305.
2. Landis, W.J. and Silver, F.H. (2002) The structure and function of normally mineralizing avian tendons, *Comparative Biochemistry and Physiology Part A* **133**, 1135–1157.
3. Bianco, P. (1992) Structure and mineralization of bone, in E. Bonucci (eds.), *Calcification in biological systems*, CRC Press, Boca Raton.
4. Diekwisch, T.G.H., Berman, B.J., Anderton, X., Gurinsky, B., Ortega, A.J., Satchell, P.G., Williams, M., Arumughan, C., Luan, X., McIntosh, J.E., Yamane, A., Carlson, D.S., Sire, J.-Y. and Shuler, C.F. (2002) Membranes, minerals and proteins of developing vertebrate enamel, *Micros. Res. Tech.* **59**, 373-395.

5. Creagh, L.T. and McDonald, M. (2003) Design and performance of inkjet print heads for non-graphic arts applications, *MRS Bulletin* **28**, 807-811.
6. Derby, B. and Reis, N. (2003) Inkjet printing of highly loaded particulate suspensions, *MRS Bulletin* **28**, 815-818.
7. Deegan, R. (2000) Pattern formation in drying drops, *Physical Review E* **61**, 475-485.
8. Shmuylovich, L., Shen, A. and Stone, H. (2002) Surface morphology of drying latex films: Multiple ring formation, *Langmuir* **18**, 3441-3445.
9. Fischer, B.J. (2002) Particle convection in an evaporating colloidal droplet, *Langmuir* **18**, 60-67.
10. Deegan, R.D., Bakajin, O., Dupont, T.F., Huber, G., Nagel, S.R. and Witten, T.A. (1997) Capillary flow as the cause of ring stains from dried liquid drops, *Nature* **389**, 827-829.
11. Shimoda, T., Mori, K., Seki, S. and Kiguchi, H. (2003) Inkjet printing of light-emitting polymer displays, *MRS Bulletin* **28**, 821-827.
12. Calvert, P. and Liu, Z. (1998) Freeform fabrication of hydrogels, *Acta Materialia* **46**, 2565-2571.

STEM CELLS AND BIOACTIVE MATERIALS

ROBERT C. BIELBY & JULIA M. POLAK
Imperial College London
Tissue Engineering and Regenerative Medicine Centre
Faculty of Medicine
Chelsea & Westminster Campus
369 Fulham Road
London, SW10 9NH
UK

1. Introduction

One of the chief goals of medicine has always been to overcome the debilitating effects of organ and tissue loss. For many centuries, removal of the diseased tissue was the only option. Greater understanding of how organs function led to the realisation of how, in some situations, a synthetic replacement might be used. Other advances in areas such as antiseptics, antibiotics and improved hygiene have all contributed to a major increase in human survivability, leading to a greater need for replacement of tissues. Limb prosthetics was the first area to make use of materials in a way which substantially advanced patient care [1]. Since then, medical implants have radiated into a staggering variety of numbers and designs [2]. The use of synthetic implants has had a major impact on healthcare. Millions of patients have had their quality of life markedly enhanced by the development and deployment in the clinical setting of implants such as total joint prostheses, cardiovascular stents and artificial heart valves. The socio-economic costs of treating tissue loss and organ failure in an increasingly ageing population (predominantly in the Western world) are vast [3] and, while artificial implants have significantly improved the quality of life for many patients, so far it has not been possible to overcome their major failing, namely limited lifespan. Implants which do not integrate with the host tissue are subject to wear and thus eventually fail, implant survivability usually being no greater than 15 years [4]. The basis of this problem lies in the selection of materials for use in implants. Implants constructed from bio-inert materials have proved to be mechanically durable but often unsatisfactory in terms of their biocompatibility, sometimes causing inflammatory responses which accelerated implant wear [5]. Improving biocompatibility has therefore been perhaps the biggest driver of innovation in medical device design in recent years. The number of "ideal" new materials being proposed for medical devices and their complexity is continuing to grow each year, as researchers around the world continue in their quest for

R.L. Reis and S. Weiner (eds.),
Learning from Nature How to Design New Implantable Biomaterials, 181-198.
© 2004 *Kluwer Academic Publishers. Printed in the Netherlands.*

improvement [6]. However, these modifications can lead to a compromise on the engineering aspects of the implant in order to improve its biocompatibility. Therefore, while it may become less susceptible to wear or failure as a result of biological processes, it might have sub-optimal mechanical properties which lead to shorter implant lifetimes. The challenge now facing biomedical science is one of its own making - namely how to address the progressive increase in patient longevity. The effect of increased patient longevity is twofold – many more patients now require these treatments and the treatments themselves are required to have longer lifetimes as well. A shift in emphasis is therefore needed, from replacement of tissues to regeneration of tissues [7].

One of the fundamental properties of living tissue in a multi-cellular organism is its capacity to adapt and remodel to physiological and environmental cues. One of the shortcomings observed in the use of synthetic implants to replace damaged or diseased tissues and organs is the inability to perceive the local conditions and respond in an appropriate fashion. A failure to adapt to the local tissue environment can be a major factor in implant failure. The next step in developing clinical implants is therefore to use bioactive materials which can either elicit a regenerative response at the site of damage *in vivo* or be used to grow tissue *in vitro* for subsequent implantation. In order to achieve this, materials must have sophisticated properties that go beyond the basics of enhancing cell adhesion and minimising inflammatory responses. The capacity to bond to and integrate with the host tissue, to alter local gene expression in a quantifiable manner, to resorb over time if necessary, to promote angiogenesis, and to be tailored at the molecular level to suit requirements are all necessary features of the new generation of bioactive materials. Combining these materials with cells will allow the fabrication of living tissue implants which have full biological function, the ability to respond to environmental changes and which possess a significantly longer lifespan than current implant devices.

2. Clinical Needs in the 21st Century

The changes in lifespan. population demographics and lifestyle are leading to a different set of clinical challenges in the 21st century to those faced previously. The prevalence of some conditions is going to increase and dealing with these is likely to be beyond the scope of current therapies, both in terms of expense and long term patient survivability.

2.1. DIABETES

Diabetes is the fifth leading cause of death by disease in the US. Current estimates are that 18 million people in the US (6.3% of the population) have diabetes [8]. The predictions for future cases make estimates as high as one in three will go on to develop the disease [9]. Healthcare costs associated with diabetes (direct medical expenditure and costs attributable to diabetes) in the US in 2002 were estimated at $132 billion [10]. Not only diabetes itself, but the complications that arise out of the disease are going to have an increasing impact on human health. While novel techniques such as islet transplantation are opening new avenues for treatment [11], the limitations of transplanted tissue are well known so there are areas of concern that must be dealt with before this can be considered a generally available treatment for diabetes.

2.2. CARDIOVASCULAR DISEASE

Cardiovascular disease, which includes coronary heart disease, stroke and high blood pressure, is the cause of 40% of all deaths in the US and approximately 62 million people in the US suffer from some form of cardiovascular disease (estimated figures from 2000 [12]). Coronary heart disease is the leading cause of death in the US and many other Western countries. While a greater range of treatment options, both surgical and pharmacological, have become available and have improved outcomes for patients, the tissues that die as a result of myocardial infarction or stroke cannot be replaced at present.

2.3. NEURODEGENERATIVE DISEASES

The degeneration of neural tissue in the brain leads to a number of diseases, Parkinson's and Alzheimer's being among them. Approximately 4.5 million people in the US suffer from Alzheimer's [13] and at least 500,000 from Parkinson's [14]. Because the incidence of these diseases increases with age, neurodegenerative disorders are also becoming much more prevalent. The impact of these conditions upon quality of life is profound and the effects can be very distressing both for the sufferer and their families. Although some drug treatments have proven effective, they are not suitable for all patients either in terms of clinical efficacy or the long term costs.

2.4. JOINT REPLACEMENT

One of the greatest areas of clinical success in tissue replacement of the past 30 years has been the use of artificial joint replacements in orthopaedic surgery. Hip replacements have been clinically successful since their introduction and knee replacements, though having many problems initially, have also led to remarkable increases in the quality of life of patients who have received them. Over 475,000 hip and knee replacements are carried out in the US every year at a cost of more than $10 billion [15]. Unfortunately, a plateau in the longevity of implant survivability seems to have been reached at around 15 years [16]. In order to cope with future clinical demands, new approaches are required to attain implant survivability in the range of 20-30 years. In particular, the challenge is to find a bioactive material or tissue regeneration therapy that behaves in a manner equivalent to an autograft, which is the gold standard in bone regeneration for orthopaedic surgeons.

3. Cells for Tissue Regeneration

One of the key choices for *in vitro* formation of tissue is the type of cell used as the starting point. The merits, potential and drawbacks of various types of stem cell are currently the topic of much research, discussion and speculation [17]. The existence and nature of stem cell niches, tissue-specific progenitors and the role of fusion are all controversial issues that remain to be resolved. Autologous cells recovered from tissue biopsies have already been used to generate tissue for implantation into cartilage defects [18]. Whether one cell source will come to predominate as a preferred solution to many types of tissue repair or whether a number of treatment styles will develop, each utilising different cells, is an open question.

3.1. EMBRYONIC STEM CELLS

The identification and isolation of embryonic stem (ES) cells from the mouse was a major discovery in biology [19, 20]. The initial enthusiasm by researchers to use ES cells in generating transgenic and knockout mice meant that the potential of ES cells in regenerative medicine was overlooked for some time after their initial discovery.

From the early Fifties, the work of Stevens and Pierce (reviewed by Alexandre [21]) demonstrated that teratocarcinomas contained cells that had multi-lineage potential. The isolation and culture of embryonal carcinoma (EC) cells gave developmental biologists an *in vitro* model in which to study the processes of differentiation [22]. The isolation of ES cells from first primate [23] and then human blastocysts [24] was the step that put them at the forefront of regenerative medicine research. Human ES cells, with the ability to proliferate (apparently) indefinitely *in vitro* and the capacity to differentiate into any somatic cell type, are clearly an important potential cell source [25]. Papers describing the differentiation of human ES cells to neural cell types (neurons, oligodendrocytes and glia), cardiomyocytes, beta cells, osteoblasts, hepatocytes and haemaotpoetic progenitors that have appeared in the short time since human ES cells were first isolated are indicative of the huge promise in the field [26-32]. The restricted numbers of human ES lines currently available, restrictions in some countries on human ES cell research and the continuing ethical and debate (particularly with regard to nuclear transfer or "therapeutic cloning") are all issues which will strongly influence possible future therapeutic use.

3.2. FETAL STEM CELLS

Wounding *in utero* is healed without scar tissue formation and fetal tissues contain immature cells with greater capacity for proliferation than adult cells. By implication, fetal cells may therefore have the ability to regenerate tissues if transplanted into adults. The use of fetal tissue transplantation has been most extensively studied in relation to neural repair in the brain and spinal cord. Transplantation of fetal dopaminergic neurons from the mesencephalon into human subjects with Parkinson's disease has produced therapeutic benefit to patients and studies showed that the transplanted cells could survive and release endogenous dopamine, thereby substituting for the specialised cells destroyed in the disease process [33-35]. While some variation was expected between patients treated at different centres, of more concern has been the variable outcomes between patients treated at the same centre. Some patients showed very little or no benefit, while others showed substantial improvement in motor function. Imaging of F-dopa uptake in the brain found highly variable results, which suggested that one of the reasons for the large amount of variability might be differences in graft survival and growth [34]. As was widely reported in the media, some of the patients suffered severe side effects, though recent data has raised the possibility that these are due to dorsal-ventral imbalance in the dopaminergic activity of the graft, rather than overgrowth of the grafted neurons as was initially suspected [36]. Greater control over the delivery of cells into areas of the striatum where they will function correctly may lead to greater consistency in the results obtained and reduce the incidence of side effects. However, as each individual treatment requires cell harvesting from several abortuses and the ethical issues associated with fetal cell transplantation are evidently complex and generate strong debate, the potential for widespread use of this therapy is likely to be somewhat limited.

The recent identification of human fetal mesenchymal stem cells (MSC) raises the possibility of using an alternative, less controversial, source of primitive cells for therapy, perhaps even permitting the harvest of autologous cells for *in utero* treatments [37]. The MSC population extracted from fetal blood contains adherent cells which divide in culture for 20 to 40 passages and can differentiate into mesenchymal lineages including bone and cartilage, but also have the ability to form oligodendrocytes and hemopoietic cells (N Fisk, pers. comm.) These cells, which can be found circulating only during the first trimester, are similar to haematopoietic populations in fetal liver and bone marrow and will engraft into multiple organs and undergo site-specific tissue differentiation when transplanted into a xenogeneic sheep model [38].

3.3. ADULT STEM CELLS

The ability of some tissues in the adult (e.g skin, haematopoietic system, bone and liver) to repair or renew indicates the presence of stem or progenitor cells. The use of autologous or allogeneic cells harvested from adult patients might provide a less difficult route to regenerative cell therapies. Human bone marrow is already transplanted into patients with haematological disorders or who are undergoing chemo- or radio-therapy for malignancies with proven therapeutic effects. It was assumed in the past that many tissue were incapable of self-regeneration if damaged because they did not possess an endogenous stem cell but, more recently, it has been discovered that many more adult tissues than was previously thought harbour cells with the capacity for regenerative repair.

3.3.1. *Tissue-Specific Stem Cells*
A number of adult tissues can mount a repair response after injury. These responses are founded upon the proliferation and differentiation of local populations of stem cells. The examples discussed here, which are by no means exhaustive, are intended to highlight some recent significant findings in this area of research.

Mesenchymal stem cells. The potential of cells found in the stroma of bone marrow to give rise to multiple non-haemopoietic lineages is well documented [39]. These cells, known as mesenchymal stem cells (MSC) or marrow stromal cells, can be purified and propagated clonally *in vitro* and have been shown to give rise to osteoblasts, chondrocytes and adipocytes depending upon the growth factors that are used to stimulate them [40]. There are obvious possibilities for using autologous MSC's in cell replacement therapy with no issues of rejection, and much of the current work is attempting to find ways to maximise the yield of these cells from relatively small amounts of bone marrow aspirate and facilitate proliferation and differerentiation *in vitro*. The possibility that stem cells from the bone marrow enter the circulation and migrate to other sites in the body where they can differentiate in order to repair tissue damage is another area of great research interest. Experiments in animal models have shown engraftment of donor cells into multiple tissues [41]. Evidence from transplant studies also suggests that this mechanism operates in humans [42], which has led to small-scale clinical trials such as TOPCARE (see section 4.1.) to assess the safety of using autologous bone marrow in treating myocardial infarction. Future treatments may utilise cytokine treatment (e.g. GM-CSF) to enhance mobilisation of cells out of the marrow space and into circulation, an approach that has shown promising results in mouse models [43].

Hepatic stem cells. The principal cell of the liver is the hepatocyte. Normally these cells are non-proliferative, but in response to cell loss they will enter the cell cycle and undergo rapid self-renewal in order to regenerate tissue. Some of this expansion in cell numbers is the result of clonal expansion, as shown by studies of hepatocyte transplantation [44]. The hepatocyte can therefore be regarded as a functional stem cell for the liver. More severe damage or blockage of normal hepatocyte regeneration after injury activates a second regenerative programme within the liver. Cells from the intrahepatic biliary tree proliferate and give rise to bipotential oval cells that differentiate into both new hepatocytes and biliary cells [45].

Pancreatic stem cells. Animal studies of pancreatic regeneration in rat have shown massive regeneration following partial pancreatectomy, including new islet formation [46]. The identification of progenitor cells responsible for this response has been complicated by the observed proliferative events in this system. Actively dividing cells only become visible 20 hours after pancreatectomy, but new islets are formed by 72 hours. Therefore, considering cell doubling times and the cell numbers in each islet, the new islets cannot be derived from a single stem cell, but rather each must contain cells derived from multiple progenitors. One model for how this might come about is the loss of mature phenotype by ductal cells which then differentiate to form islets. When cultured *in vitro* under the right conditions, adult ductal cells can be induced to differentiate into glucose-responsive beta cells [47]. The homeodomain transcription factor PDX-1, which is essential for normal pancreatic development, has been shown to be able to induce insulin production in cultures of ductal cells, providing further evidence in favour of the hypothesis that duct cells act as progenitors in the pancreas [48, 49].

3.3.2. Multipotent Adult Stem Cells

The accepted view of adult stem cells for most of the time they have been studied in tissues such as skin and bone marrow was that their differentiation potential was strictly limited to cell lineages found within the tissue of origin. In the last few years this view has been challenged. Several reports have claimed to show *plasticity* of adult stem cells i.e. the ability to differentiate to cell types different from the tissue of origin. A comprehensive account of this emerging and controversial field is not possible here, but an excellent review by Poulsom and co-workers [50] has recently been published.

Side population (SP) cells. Small fractions of the bone marrow contain a population of cells that can be identified by their ability to exclude the fluorescent dye Hoechst 33342 [51]. When purified via fluorescent activated cell sorting (FACS), the cells appear as a distinct population on the side of the fluorescent profile, hence their designation as 'SP'. The characteristic SP phenotype appears to be associated with the expression of a protein called the ABCG2 transporter [52]. In damaged liver, several members of the ABC protein family of membrane transporters were upregulated in regenerating ductules, the proposed site of hepatocyte stem cell compartment ([53] see also section 3.3.1.) The identification of SP cells from a number of tissues and the expression of members of the ABCG2 transporter on a number of types of stem cell has led to the suggestion that it may be a determinant of the side population phenotype [54].

Multipotent adult stem cells. Work by Verfaillie and colleagues identified a small pool of cells that co-purify with mesenchymal stem cells (MSC) from bone marrow but have much greater differentiation potential [55, 56]. Termed multipotent adult progenitor cells (MAPC), they were differentiated *in vitro* to cells of the mesodermal, endodermal and neurectodermal lineages. This remarkable finding challenged long held assumptions that truly multi- or pluripotent stem cells in the adult did not persist beyond early stages in embryogenesis. MAPC express oct-4 and rex-1, markers of pluripotency previously only seen in embryonic stem cells or the pre-gastrulation embryo and can contribute to multiple lineages when injected into a blastocyst. A number of questions remain to be answered: do these cells exist and elaborate this multi-lineage potential *in vivo*, or are they an artefact of culture? Are they the same as side population cells? What is certain is that these unexpected findings have stimulated much greater interest in the possibilities for plasticity in adult cells.

3.4. TERMINALLY DIFFERENTIATED CELLS

The clinical success of transplantation in a wide variety of organs and tissues has shown that terminally differentiated adult cells are a therapeutic alternative provided that they can be made available in sufficient numbers. Transplants of hepatocytes, pancreatic islets (the Edmonton protocol for treatment of diabetes) and autologous chondrocyte implantation (ACI, e.g. Carticell®) for articular cartilage lesions are all in use for patient treatment. The challenge is to obtain enough cells to restore function to the damaged tissue. In the case of ACI, cells are obtained from a small biopsy and expanded *in vitro* before re-implantation. This poses a problem in the use of chondrocytes, as it is well known that prolonged 2-dimensional culture in monolayer causes them to de-differentiate [57]. This loss of phenotype can be reversed by a period of 3-dimensional culture following encapsulation [58]. A number of other terminally differentiated cells are also notoriously difficult to maintain in culture, which has partly led to the increased interest in fetal cells and stem cells as alternative sources.

4. *In Situ* Tissue Regeneration

4.1. TISSUE REGENERATION THROUGH CELL REPLACEMENT

One of the possible strategies for addressing the predicted clinical needs of the future is cell replacement therapy. Replacement of cells *en masse* via an organ or tissue transplant is nothing new – Christian Barnard performed the first heart transplant in 1967 [59]. However, the acute shortage of donor organs available for transplantation means that it is an option for the few, not the many. The sheer numbers of patients requiring treatment will mean that organ transplantation in its current form will never be able to meet demand. Therefore, cell replacement therapy is being explored with a great deal of interest. This approach utilises *ex vivo* expansion or selection of stem or progenitor cells which are then transplanted back into patients where they restore structure, function and the capacity for future adaptation to changing physiological parameters. Bone marrow transplants have been used in this way for a number of years to treat haematological disorders and, in one notable study, children with osteogenesis imperfecta were treated with allogeneic transplants of marrow-derived mesenchymal progenitor cells from healthy individuals which resulted in an improvement in their conditions [60]. Studies in rodents have strongly suggested that bone marrow-derived progenitors are able to migrate to damaged cardiac muscle tissue and improve cardiac function following experimentally induced infarction [61, 62], although attempts to repeat this work in non-human primates have so far failed to yield any positive results [63]. A small clinical study (TOPCARE) is currently in progress to examine the potential therapeutic benefit of infusing autologous bone marrow into patients who have recently suffered myocardial infarction [64]. Initial follow-up has been promising, but it is too early at this stage to draw definitive conclusions about the benefits or potential risks of this approach [65].

4.2. THE ROLE OF BIOACTIVE MATERIALS IN CELL REPLACEMENT

The potential clinical impact of cell replacement therapy is immense and the use of bioactive materials will play a central role in bringing this about. Surface modifications of implant materials at the molecular level have been designed to promote cell adhesion and tissue bonding so that a better interface between the host tissue and the synthetic implant is created. It has been hoped that research and development work into bioactive fixation, improving the interfacial bonding between implant and tissue, would secure an improvement in implant survivability. Unfortunately, bioactive fixation has not brought about the gains that had been hoped [16]. This is probably principally due to mechanical imbalances at the interface and the inability to remodel in response to applied load. For cell replacement to be most effective and also safe, a structure is needed on which the cells can be seeded, grown and (re)generate tissue which is capable of further adaptation to the physiological and mechanical environment. This might either be for *in vitro* growth of tissue constructs prior to implantation or to provide anchorage sites and mechanical stability once the cells are implanted. There is the further benefit of providing biological cues through controlled release of biologically active molecules which have been incorporated into the material. Materials can now be synthesised with a hierarchy of textural features, ranging from the nanometer to the millimetre scale [66,67]. Inorganic and inorganic-organic composite materials can be produced with the ability to control the rates of resorbtion, the surface chemistry and release of inorganic or organic species, enabling controlled stimulation of cell proliferation and differentiation [68].

4.2.1. *Activation of Local Gene Expression*
Specification and morphogenesis of tissue requires that cells undergo a correct temporal and spatial pattern of changes in gene expression, leading to elaboration of the fully differentiated phenotype. This is demonstrated very readily by cultures of primary human osteoblasts, where phases of proliferation, matrix synthesis and biomineralisation all occur in sequence because of the underlying pattern of changes in gene expression [69]. In order to achieve successful regeneration of tissues, cells must receive stimuli that promote the correct pattern of gene expression events. These must

also be localised as far as possible to the area of damage or tissue loss. Studies on 45S5 Bioglass[®] and primary human osteoblasts have demonstrated the activation of cell proliferation, IGF-II secretion and up-regulation of a number of genes involved in osteogenesis [70-72]. Significantly, this stimulation does not require impregnation of the material with any growth factors: the hydroxyapatite surface formed on the ceramic and the ions released during resorbtion are sufficient to trigger the sequence of gene expression changes.

Triggering gene expression and bone tissue formation can also be achieved using growth factors such as bone morphogenetic protein 2 (BMP-2) [73]. Several studies have looked at methods for delivering this molecule from biomaterial scaffolds [74-77]. Polymers have been widely used as delivery vehicles because they are biocompatible, degrade in a controlled manner *in vivo* and can be easily processed into three-dimensional porous structures. The range of degradation rates and porosities possible using polymer materials such as poly(lactic acid) can provide a controlled release delivery system for physiological, local doses of growth factors. Encapsulation of BMP-2 in microspheres has been shown to be an effective way of delivering BMP-2 *in vitro* and *in vivo* where it promotes cortical bone formation in a radial defect model [76]. More recently, supercritical fluid methods have been used to generate porous, biodegradable polymer scaffolds containing loaded with protein [78,79]. The absence of either solvents or the need for heating makes this a highly promising route to incorporating growth factors such as BMP-2 into materials without compromising its biological activity.

4.2.2. *Angiogenesis*

Achieving adequate vascular supply of any tissue implant, whether it is grown *ex vivo* or generated *in situ* by cell implantation, is one of the major limitations of tissue regeneration strategies. One of the obstacles to be overcome is ensuring sufficient porosity and interconnectivity of pores within the scaffold to permit vascular invasion and development of a vascular network. In the past this could be achieved by addition of a heat or solvent labile component such as PEG during scaffold synthesis which can be subsequently removed prior to cell seeding [80]. More advanced processing techniques such as foaming during sol-gel ceramic synthesis [81] or supercritical fluid methods [78] create pores of a range of sizes in the material during fabrication.

Another way of promoting vascularisation is by the incorporation of biological agents that stimulate angiogenesis such as endothelial growth factor (VEGF). Hydrogels have been used successfully to encapsulate VEGF while retaining its biological activity. Implantation of alginate hydrogels with and without VEGF in SCID mice has been shown to enhance blood vessel formation by the release of VEGF from the loaded scaffolds [82,83]. Polyethylene glycol based hydrogels have also been used to carry VEGF. By incorporating peptide sequences that mediate cell adhesion (RGD) and cross-linking the gel using peptides which were susceptible to degradation by matrix metalloproteinases (CPQG-IWGQ), it was possible to make VEGF available to cells locally as they invaded and remodelled the material [84]. Although the mechanical properties of hydrogels limits their application, these new avenues in imparting angiogenic bioactivity to scaffold matrices are important steps towards ensuring that regeneration of large pieces of tissue becomes a clinical possibility.

In the future, new types of bioactive materials may also be developed which are able to stimulate local release of angiogenic growth factors from surrounding vasculature.

Animal studies of regeneration of small intestine using bioactive ceramics have provided initial observations that endogenous release of VEGF can be induced in response to the material which results in efficient vascularisation of implanted tissue (unpublished data).

4.2.3. *Cell Guidance*

Implantation of cells *in situ* at the site of tissue loss, even with high levels of stable engraftment and survival, may not be sufficient to regenerate function. Guidance of cell polarity, growth and/or migration may be necessary. For example, in spinal cord injury, axonal regeneration across the lesion is necessary to restore nerve connections via a "bridge" which spans the gap. In such cases, bioactive materials which can provide directional cues to cells are essential to generating a functional repair. Synthetic poly(L-lactide) filaments coated with extracellular matrix proteins, such as laminin can support cell growth and direct cell orientation [85]. Scaffolds carrying cells (particularly Schwann cells to assist with myelination of axons) or releasing neurotrophic growth factors have produced very promising results in animal models (reviewed by Novikova *et al.* [86]).

5. *In Vitro* Growth of Tissue

In addition to (re)implanting cells and/or bioactive materials at the site of defects in order to stimulate *in situ* regeneration, the engineering of large pieces of tissue *in vitro* for subsequent implantation is an alternative strategy for tissue regeneration. There are a number of elements which must be brought together to enable this type of strategy to work: i) a source of cells (already discussed in section 3) ii) the ability to control the expansion and differentiation of the cells *in vitro* iii) a suitable scaffold on which to form tissue iv) control of the *in vitro* culture environment, (which will not be discussed in this paper, but see recent review by Vunjak-Novakovich [87]).

5.1. CELL GROWTH AND DIFFERENTIATION *IN VITRO*

The wealth of data accumulated through years of *in vitro* culture of primary cells has been an invaluable resource for researchers seeking to direct differentiation of stem or progenitor cells. A wide variety of cell types have been successfully derived through *in vitro* differentiation [25], though few have yet been used therapeutically. Nonetheless, studies to date show great promise in providing those cell types which will be most important in novel therapies to the diseases identified as being the major clinical challenges in the coming century (see sections 2.1 – 2.4).

5.1.1. *Cell Senescence* In Vitro

Replicative senescence is not a recent concept. In contrast to established tumour cell lines such as HeLa, most cells are not immortal in culture. It was suggested over forty years ago that normal human cells had a limit on the number of divisions that they could undergo [88]. The discovery of telomeres (tandem repeats of non-coding DNA found at the end of chromosomes) and their progressive shortening with replication and age suggested an explanation for cell senescence [89-91] which was confirmed by over-expression of telomerase (an enzyme that maintains telomere length) in normal human fibroblasts which caused their immortalization [92,93]. The limit on the number of rounds of division a cell can undergo is a severe obstacle in producing large banks of high quality cells for therapy. While there are molecular mechanisms for over- or re-expressing telomerase in cells, there are obvious concerns about tumorigenicity of immortalised cells post-implantation. Mechanisms for stopping growth or killing the cells in the event of uncontrolled growth are therefore necessary. One way is to use the telomerase (hTERT) promoter to drive the expression of molecules that would ensure, in the event of cells becoming tumorigenic, that they are removed. Such a method has been developed in human ES cells at the Roslin Institute (J. McWhir, pers. comm.) and utilises the xenotransplantation antigen gal 1,3. A functional galactosyl transferase (GalT) gene is placed under the control of the hTERT promoter. The GalT can generate gal 1,3 epitopes on the cell surface, making the cells susceptible to complement-mediated cell lysis in human serum. However, even considering the advances made in this area, genetic manipulation of cells prior to transplantation will have to satisfy a large number of regulatory questions before becoming acceptable for clinical use.

5.1.2. Cell Differentiation In Vitro
Expression of stable phenotypes by primary cells in culture has been studied for many years in order to create useful *in vitro* models of tissue physiology. Combinations of growth factors and specialised media formulations for many cell types are well documented and widely used. Controlling differentiation, particularly of stem cells, *in vitro* has drawn heavily on this background of knowledge and also from developmental biology studies of mechanisms of cell specification and differentiation.

Chondrocytes. The difficulties of maintaining the chondrocyte phenotype *in vitro* have already been briefly mentioned. Although it is relatively easy to extract viable chondrocytes from cartilage samples by collagenase digestion [94], the chondrocyte has a propensity to de-differentiate into a fibroblast phenotype after prolonged culture (i.e. multiple passages) *in vitro*. A plethora of factors and growth conditions have been used to maintain or restore expression of type II collagen and other principal chondrocyte phenotype markers. These include members of the TGF-ß growth factor super-family [95], 3-dimensional culture [96], compressive loading [97] and oxygen tension [98,99]. Sophisticated bioreactor designs are now being evolved to enable *in vitro* growth of large pieces of tissue suitable for implantation [87].

Osteoblasts. Isolation of mineralising cells from mammalian bone was achieved over 30 years ago [100], and discovery of the necessary conditions to induce the osteoblast to undergo its normal differentiation program *in vitro* followed later [101]. A combination of ascorbic acid and a source of inorganic phosphate (sodium beta glycerophosphate is most commonly used) are a minimum requirement, although the supra-physiological doses used in many studies still possibly limit the extension of findings to the *in vivo* situation. The synthetic glucocorticoid hormone dexamethasone [102] as well as vitamin D3 [103] are often also added to osteoblast cultures as well as other stimulatory growth factors such as BMPs [104]. Once reliable methods for instructing osteoblasts to differentiated in culture were established, the patterns of cell proliferation and gene expression have been carefully studied [105]. This model system is now widely used to study mechanisms of tissue-specific transcriptional control e.g. of the osteocalcin gene [106], but these studies are invaluable in establishing reliable protocols for controlling proliferation and full elaboration of the mature phenotype *in vitro*.

5.2. USE OF BIOACTIVE MATERIALS *IN VITRO*

In vitro testing of bioactive materials prior to use in animal models and, perhaps eventually, in a clinical setting is *de rigeur* in modern research, and is facilitated by the increased availability of suitable cells for testing and greater inter-disciplinary research. Previously, biomaterials and the implants constructed from them were developed through trial and error experimentation [2]. Novel materials are now exposed to cultured cells to evaluate their biocompatibility, toxicity and bioactivity, although this has yet to remove the need for testing in animal models. The literature dealing with biomaterials and their interactions with cells has expanded at an astonishingly rapid rate in recent years. There are numerous uses of bioactive materials *in vitro* beyond straightforward routine testing. Anchorage and cell-substrate interactions are clearly important primary considerations, as can be seen by reference to propagation of ES cells. Currently, most human ES cell lines are grown on layers of feeder cells – either mouse or human – and this methodology is one of the obstacles to scale-up of production for therapeutic use. Feeder-free growth has been achieved for short periods using either Matrigel® or recombinant human laminin [107]. However, both of these are prohibitively expensive for industrial scale culture, so a synthetic alternative that could mimic their attachment properties for hES culture would have a significant impact on the translation of research to clinical usage. Incorporating functional domains of proteins onto materials can make a multi-functional surface or tailor them for cell-specific recognition. Coating of sol-gel bioactive ceramic with laminin for example, causes greater adhesion of lung epithelial cells than on the ceramic alone [108].

One of the key features of many new biomaterials is their ability to directly alter changes in cellular metabolism and gene expression, as opposed to solely being an anchorage site. The combination of resorbable and bioactive properties into composites is giving rise to a whole new class of biomaterials – the "third generation" that activate cells and genes to elicit a particular biological response [68]. Bioactive scaffolds can be designed to provide optimal attachment topography while also releasing optimal dosages of inorganic ions into the local environment of the cell (Fig 1 & 2).

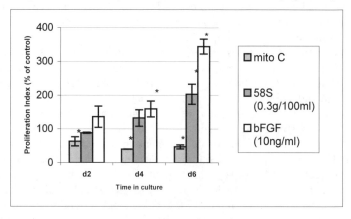

Figure 1. Effect of soluble inorganic ions released by dissolution of 58S sol-gel bioactive glass on proliferation of primary human osteoblasts compared to a known mitogen (10ng/ml basic FGF) or a mitotic inhibitor (Mitomycin C). Mean +/- s.e.m.; * $p < 0.05$ compared to control.

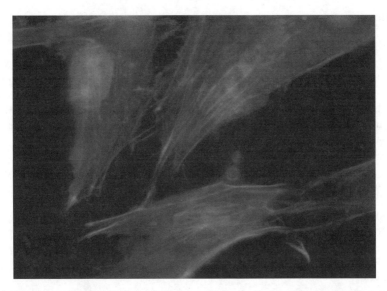

Figure 2. Fluorescent staining (green) of actin filament organization in primary human osteoblasts following 24 hours of attachment on a composite 95:5 PDLLA/Bioglass disc, with propidium iodide nuclear counterstain (red). (Courtesy of Olga Tsigkou, Imperial College London.)

Amongst the available biodegradable polymers, poly(lactic acid) (PLA), poly(glycolic acid) (PGA) and poly (DL-lactic-co-glycolic acid) (PLGA) have advantages in being FDA approved and also currently in use in patients. PLGA scaffolds have been modified to act as controlled release vehicles for a number of osteo-stimulatory molecules such as ascorbate-2-phosphate, dexamethasone [109] and BMP-2 [76] as well as other biologically active compounds such as nitric oxide [110] and insulin [111].

Three-dimensional PLGA scaffolds have been modified to carry the RGD ligand on their surface, thereby encouraging cell attachment through integrin adhesion [112]. Three- dimensional culture on scaffolds *in vitro* also holds some significant advantages over traditional two-dimensional methods. There are numerous studies where three-dimensional culture of cells on bioactive/biomaterial scaffolds have had success in producing organised tissues. Human embryonic stem cells grown on 50:50 PLLA/PLGA sponges were found to differentiate and organise into structures with recognisable features of committed tissues [113]. Stimulation with retinoic acid, TGF-beta, activin A or IGF induced differentiation of the cells to form three dimensional structures with features of neural, cartilage and liver tissues respectively. By contrast, growth on fibronectin coated two-dimensional culture dishes failed to support formation of three-dimensional structures. The tissue constructs formed in this study were implanted *in vivo* into SCID mice, where the cells remained viable and the tissue became vascularised by the host.

The complexity of scaffolds now being produced, which allows optimal cell attachment, controlled release of stimulatory factors and resorbtion matched to tissue formation rates, holds the great promise of being able to design specific, gene-activating, regenerative materials tailored to particular diseases and possibly even specific patients.

6. Summary

Major advances in biological and materials research have created the possibilities for tissue engineering and regenerative medicine. Finding the most effective ways of utilising stem cells, of several types, and triggering their differentiation in a controlled manner will provide cell sources for cell replacement therapy. Materials will be bioresorbable *in vivo* and bioactive, contributing to differentiation, implantation and long-term engraftment of cells. Developing these concepts from bench to bedside will be crucial in meeting healthcare needs in the coming century.

7. Acknowledgements

The authors would like to thank the Medical Research Council (UK), The March of Dimes and The Rosetree Trust for supporting their work.

References

1. Sanders, G.T. (1986) *Amputation Prosthetics*, F.A. Davis Company, Philadelphia.
2. Applications of Materials in Medicine and Dentistry. Ch. 7 in Ratner, B.D., Hoffman, A.S. and Schoen, F.J., Lemmons, J.E. (eds) (1996), *Biomaterials Science*, Academic Press, San Diego, p.283-388.
3. Chapekar, M.S. (2000) Tissue Engineering: Challenges and opportunities, *J. Biomed. Mater. Res.* **53**, 617-620.
4. Spector, M. (1992) Biomaterial failure, *Orthop. Clin. North Am.* **23**, 211-217.
5. Anderson, J.M. (1994) Inflammation and the foreign body response, *Prob. Gen. Surg.*, **11**, 147.
6. Hutmacher, D.W. (2000) Scaffolds in tissue engineering bone and cartilage, *Biomaterials* **21**, 2529-2543.
7. Hench, L.L. (1998) Biomaterials: a forecast for the future, *Biomaterials* **19**, 1419-1423.
8. National Institute of Diabetes and Digestive and Kidney Diseases (2003) *National Diabetes Statistics fact sheet: general information and national estimates on diabetes in the United States*, U.S. Department of Health and Human Services, National Institutes of Health, Bethesda, MD.
9. Narayan, K.M.V., Boyle, J.P., Thompson, T.J. and Sorensen; David F. Williamson, S.W. (2003) Lifetime Risk for Diabetes Mellitus in the United States, *JAMA* **290**, 1884-1890.
10. Hogan, P., Dall, T.and Nikolov, P. (2003) Economic costs of diabetes in the US in 2002, *Diabetes Care* **26**, 917-932.
11. Ryan E.A., Lakey J.R., Rajotte R.V., Korbutt G.S., Kin T., Imes S., Rabinovitch A., Elliott J.F., Bigam D., Kneteman N.M., Warnock G.L., Larsen I. and Shapiro A.M. (2001) Clinical outcomes and insulin secretion after islet transplantation with the Edmonton protocol, *Diabetes* **50**, 710-719.
12. American Heart Association (2002) *Heart Disease and Stroke Statistics — 2003 Update*, American Heart Association, Dallas, Tex.
13. Alzheimer's Association (2003, August) *Statistics about Alzheimer's disease*, http://www.alz.org/ResourceCenter/FactSheets/FSAlzheimerStats.pdf, Accessed 2003, October.
14. National Institute of Neurological Disorders and Stroke (2003, July) *Parkinson's Disease Backgrounder*, http://www.ninds.nih.gov/health_and_medical/pubs/parkinson's_disease_backgrounder.htm, accessed 2003, October.
15. Agency for Healthcare Research and Quality, Rockville, MD (2003, July), *HCUPnet, Healthcare Cost and Utilization Project*. http://www.ahrq.gov/data/hcup/hcupnet.htm, accessed 2003, October.
16. Hench, L.L. and Wilson, J. (1996) *Clinical performance of skeletal prostheses*, Chapman and Hall, London.
17. Verfaillie, C.M., Pera, M.F. and Lansdorp, P.M. (2002) Stem cells: hype and reality, *Haematology* **2002** 369-391.
18. Mayhew, T.A., Williams, G.R., Senica, M.A., Kuniholm, G. and Du Moulin, G.C. (1998) Validation of a quality assurance program for autologous chondrocyte implantation, *Tiss. Eng.* **4**, 325-334.

19.   Evans, M.J. and Kaufman, M.H. (1981) Establishment in culture of pluripotential cells from mouse embryos, *Nature* **292**, 154-156.

20.   Martin, G.R. (1981) Isolation of a pluripotent cell line from early mouse embryos cultured in medium conditioned by teratocarcinoma stem cells *Proc. Natl. Acad. Sci. USA* **78**, 7634-7638.

21.   Alexandre, H. (2001) A history of mammalian embryological research, *Int. J. Dev. Biol.* **45**, 457-467.

22.   Martin, G.R. and Evans, M.J. (1975) Differentiation of clonal lines of teratocarcinoma cells: formation of embryoid bodies in vitro, *Proc. Natl. Acad. Sci. USA* **72**, 1441-1445.

23.   Thomson, J.A., Kalishman, J., Golos, T.G., Durning, M., Harris, C.P. Becker, R.A. and Hearn, J.P. (1995) Isolation of a primate embryonic stem cell line, *Proc. Natl. Acad. Sci. USA* **92**, 7844-7848.

24.   Thomson J.A., Itskovitz-Eldor J., Shapiro S.S., Waknitz M.A., Swiergiel J.J., Marshall V.S. and Jones J.M. (1998) Embryonic stem cell lines derived from human blastocysts, *Science* **282**, 1145-1147.

25.   Odorico, J.S., Kaufman, D.S. and Thomson, J.A. (2001) Multilineage differentiation from human embryonic stem cells, *Stem Cells* **19**, 193-204.

26.   Zhang, S.C., Wernig, M., Duncan, I.D., Brustle, O. and Thomson, J.A. (2001) In vitro differentiation of transplantable neural precursors from human embryonic stem cells, *Nat. Biotechnol.* **19**, 1117-1118.

27.   Mummery C., Ward D., van den Brink C.E., Bird S.D., Doevendans P.A., Opthof T., Brutel de la Riviere A., Tertoolen L., van der Heyden M. and Pera M. (2002) Cardiomyocyte differentiation of mouse and human embryonic stem cells, *J. Anat.* **200**, 233-242.

28.   Assady, S., Maor, G., Amit, M., Itskovitz-Eldor, J., Skorecki, K.L. and Tzukerman, M. (2001) Insulin production by human embryonic stem cells, *Diabetes* **50**, 1691-1697.

29.   Sottile, V., Thomson, A. and McWhir, J. (2003) In vitro osteogenic differentiation of human ES cells, *Cloning Stem Cells* **5**, 149-155.

30.   Bielby, R.C., Boccaccini, A.R., Polak, J.M. and Buttery, L.D.K. In vitro differentiation and in vivo mineralization of osteogenic cells derived from human embryonic stem cells, *Tiss. Eng.* (submitted).

31.   Schuldiner, M., Yanuka, O., Itskovitz-Eldor, J., Melton, D.A. and Benvenisty, N. (2000) Effects of eight growth factors on the differentiation of cells derived from human embryonic stem cells, *Proc. Natl. Acad. Sci. USA* **97**, 11307-11312.

32.   Kaufman, D.S., Hanson, E.T., Lewis, R.T, Auerbach, R. and Thomson, J.A. (2001) Hematopoietic colony-forming cells derived from human embryonic stem cells, *Proc. Natl. Acad. Sci. USA* **98**, 10716-10721.

33.   Freed, C.R., Greene, P.E., Breeze, R.E. Tsai, W.Y., DuMouchel, W., Kao, R., Dillon, S., Winfield, H., Culver, S., Trojanowski, J.Q., Eidelberg D. and Fahn, S. (2001) Transplantation of embryonic dopamine neurons for severe Parkinson's disease, *N. Engl. J. Med.* **344**, 710-719.

34.   Dunnett, S.B., Björklund, A. and Lindvall, O. (2001) Cell therapy in Parkinson's disease: stop or go? *Nat. Rev. Neurosci.* **2**, 365-369.

35.   Lindvall, O. (2001) Parkinsons disease: stem cell transplantation, *Lancet* **358**, S48.

36.   Ma, Y., Feigin, A., Dhawan, V., Fukuda, M., Shi, Q., Greene, P., Breeze, R., Fahn, S., Freed, C. and Eidelberg, D. (2002) Dyskinesia after fetal cell transplantation for parkinsonism: a PET study, *Ann. Neurol.* **52**, 628-634.

37.   Campagnoli, C., Roberts, I.A., Kumar, S., Bennett, P.R., Bellantuono, I. and Fisk, N.M. (2001) Identification of mesenchymal stem/progenitor cells in human first-trimester fetal blood, liver, and bone marrow, *Blood* **98**, 2396-2402.

38.   MacKenzie, T.S., Campagnoli, C., Almeida-Porada, G., Fisk, N.M. and Flake, A.W. (2001) Circulating human fetal stromal cells engraft and differentiate into multiple tissues following transplantation into pre-immune fetal lambs, *Blood* **98**, 238a.

39.   Caplan, A.I. and Bruder, S.P. (2001) Mesenchymal stem cells: building blocks for molecular medicine in the 21st century, *Trends Mol. Med.* **7**, 259-264.

40.   Pittenger, M.F., Mackay, A.M., Beck, S.C., Jaiswal, R.K., Douglas, R., Mosca, J.D., Moorman, M.A., Simonetti, D.W., Craig, S., Marshak, D.R. (1999) Multilineage potential of adult human mesenchymal stem cells, *Science* **284**, 143-7

41.   Liechty, K.W., MacKenzie, T.C., Shaaban, A.F., Radu, A., Moseley, A.M., Deans, R., Marshak, D.R. and Flake, A.W. (2000) Human mesenchymal stem cells engraft and demonstrate site-specific differentiation after in utero transplantation in sheep, *Nat. Med.* **6**, 1282-1286.

42.   Alison, M.R., Pou,son, R., Jeffery, R., Dhillon, A.P., Quaglia, A., Jacob, J., Novelli, M., Prentice, G., Williamson, J. and Wright, N.A. (2000) Hepatocytes from non-hepatic adult stem cells, *Nature* **406**, 257.

43.   Orlic, D. (2003) Adult bone marrow stem cells regenerate myocardium in ischemic heart disease, *Ann. N. Y. Acad. Sci.* **996**, 152-157.

196

44. Overturf, K., al-Dhalimy, M., Ou, C.N., Finegold, M. and Grompe, M. (1997) Serial transplantation reveals the stem-cell-like regenerative potential of adult mouse hepatocytes, *Am. J. Pathol.* **151**, 1273-1280.

45. Alison, M.R. (1998) Liver stem cells: a two compartment system, *Curr. Opin. Cell Biol.* **10**, 710-715.

46. Bonner-Weir, S., Baxter, L.A., Schuppin, G.T. and Smith, F.E. (1993) A second pathway for regeneration of adult exocrine and endocrine pancreas. A possible recapitulation of embryonic development, *Diabetes* **42**, 1715-1720.

47. Bonner-Weir, S., Taneja, M., Weir, G.C., Tatarkiewicz, K., Song, K.H., Sharma, A and O'Neil, J.J. (2000) In vitro cultivation of human islets from expanded ductal tissue, *Proc. Natl. Acad. Sci. USA* **97**, 7999-8004.

48. Noguchi, H., Kaneto, H., Weir, G.C. and Bonner-Weir, S. (2003) PDX-1 protein containing its own antennapaedia-like protein like transduction domain can transduce pancreatic duct and islet cells, *Diabetes* **52**, 1732-1737.

49. Bonner-Weir, S. and Sharma, A. (2002) Pancreatic stem cells, *J. Pathol.* **197**, 519-526.

50. Poulsom, R., Alison, M.R., Forbes, S.J. and Wright, N.A. (2002) Adult stem cell plasticity, *J. Pathol.* **197**, 441-456.

51. Goddell, M.A., Brose, K., Paradis, G., Conner, A. and Mulligan, R. (1996) Isolation and functional properties of murine hematopoietic stem cells that are replicating in vivo, *J. Exp. Med.* **183**, 1797-1806.

52. Kim, M., Turnquist, H., Jackson, J., Sgagias, M., Yan, Y., Gong, M., Dean, M., Sharp, J.G. and Cowan, K. (2002) The multidrug resistance transporter ABCG2 (breast cancer resistant protein 1) effluxes Hoechst 33342 and is overexpressed in hematopoietic stem cells, *Clin. Cancer Res.* **8**, 22-28.

53. Ros, J.E., Libbrecht, L., Geuken, M., Jansen, P.L.M. and Roskams, T.A.D. (2003) High expression of MDR1, MRP1 and MRP3 in the hepatic progenitor cell compartment and hepatocytes in severe human liver disease, *J. Pathol.* **200**, 553-560.

54. Zhou, S., Schuetz, J.D., Bunting, K.D., Colapietro, A.M., Sampath, J., Morris, J.J., Lagutina, I., Grosveld, G.C., Osawa, M., Nakauchi, H. and Sorrentino, B.P. (2001) The ABC transporter Bcrp1/ABCG2 is expressed in a wide variety of stem cells and is a molecular determinant of the side-population phenotype, *Nature Med.* **7**, 1028-1034.

55. Reyes, M. and Verfaillie, C.M. (2001) Characterization of multipotent adult progenitor cells, a subpopulation of mesenchymal stem cells, *Ann. N.Y. Acad. Sci.* **938**, 231-233.

56. Jiang, Y., Jahagirdar, B.N., Reinhardt, R.L., Schwartz, R.E., Keene, C.D., Ortiz-Gonzalez, X.R., Reyes, M., Lenvik, T., Lund, T., Blackstad, M., Du, J., Aldrich, S., Lisberg, A., Low, W.C., Largaespada, D.A. and Verfaillie, C.M. (2002) Pluripotency of mesenchymal stem cells derived from adult marrow, *Nature* **418**, 41-49.

57. Grundmann, K., Zimmermann, B., Barrach, H.J, and Merker, H.J. (1980) Behvaiour of epiphyseal mouse chondrocyte populations in monolayer culture, *Virchows Arch. A. Pathol. Anat. Histol.* **389**, 167-187.

58. Benya, P.D. and Schaffer, J.D. (1982) Dedifferentiated chondrocytes reexpress the differentiated collagen phenotype when cultured in agarose gels, *Cell* **30**, 215-224.

59. Barnard, C.N. (1967) The operation. A human cardiac transplant: an interim report of a successful operation performed at Groote Schuur Hospital, Cape Town, *S. Afr. Med. J.* **41**, 1271-1274.

60. Horwitz, E.M., Prockop, D.J., Fitzpatrick, L.A., Koo, W.W., Gordon, P.L., Neel, M., Sussman, M., Orchard, P., Marx, J.C., Pyeritz, R.E. and Brenner, M.K. (1999) Transplantability and therapeutic effects of bone marrow-derived mesenchymal cells in children with osteogenesis imperfecta, *Nature Med.* **5**, 309-313.

61. Orlic, D., Kajstura, J., Chimenti, S., Limana, F., Jakoniuk, I., Quaini, F., Nadal-Ginard, B., Bodine, D.M., Leri, A. and Anversa, P. (2001) Mobilized bone marrow cells repair the infarcted heart, improving function and survival, *Proc. Natl. Acad. Sci. USA* **98**, 10344-10349.

62. Orlic, D., Kajstura, J., Chimenti, S., Jakoniuk, I., Anderson, S.M., Li, B., Pickel, J., McKay, R., Nadal-Ginard, B., Bodine, D.M., Leri, A. and Anversa, P. Bone marrow cells regenerate infarcted myocardium, *Nature* **410**, 701-705.

63. Arai, A.E., Sheikh, F., Agyeman, K.O., Hoyt, R., Sachdev, B., Yu, X.X., San, H., Metzger, M., Dunbar, C. and Orlic, D. (2003) Lack of benefit from cytokine mobilized stem cell therapy for acute myocardial infarction in nonhuman primates, *J. Am. Coll. Cardiol.* **41(Suppl. B)**, 371.

64. Assmus, B., Schachinger, V., Teupe, C., Britten, M., Lehmann, R., Dobert, N., Grunwald, F., Aicher, A., Urbich, C., Martin, H., Hoelzer, D., Dimmeler, S.and Zeiher, A.M. (2002) Transplantation of progenitor cells and regeneration enhancement in acute myocardial infarction (TOPCARE-AMI), *Circulation* **106**, 3009-3017.

65. Schachinger, V., Assmus, B., Lehmann, R., Teupe, C., Britten, M., Doebert, N., Martin, H., Dimmeler, S., Zeiher, A.M. and Goethe, J.W. (2003) Clinical evidence for benefit of intracoronary progenitor cell

therapy on post-infarction remodeling: Results of the TOPCARE-AMI trial, *J. Am. Coll. Cardiol.* **41 (Suppl. B)**, 404.

66. Livage, J. (1997) Sol-gel processes, *Curr. Opin. Solid State Mater.* **2**, 132-138.
67. Hench, L.L. (1997) Sol-gel materials for bioceramic applications, *Curr. Opin. Solid State Mater.* **2**, 604-606.
68. Hench, L.L. and Polak, J.M. (2002) Third-generation biomedical materials, *Science* **295**, 1014-1017.
69. Lian, J.B., Stein, G.S., Stein, J.L. and van Wijnen, A.J. (1998) Transcriptional control of osteoblast differentiation, *Biochem. Soc. Trans.* **26**, 14-21.
70. Xynos, I.D., Hukkanen, M.V., Batten, J.J., Buttery, L.D., Hench, L.L. and Polak, J.M. (2000) Bioglass 45S5 stimulates osteoblast turnover and enhances bone formation In vitro: implications and applications for bone tissue engineering, *Calcif. Tiss. Int.* **67**, 321-329.
71. Xynos, I.D., Edgar, A.J., Buttery, L.D., Hench, L.L. and Polak, J.M. (2000) Ionic products of bioactive glass dissolution increase proliferation of human osteoblasts and induce insulin-like growth factor II mRNA expression and protein synthesis, *Biochem. Biophys. Res. Comm.* **276**, 461-465.
72. Xynos, I.D., Edgar, A.J., Buttery, L.D., Hench, L.L. and Polak, J.M. (2001) Gene-expression profiling of human osteoblasts following treatment with the ionic products of Bioglass 45S5 dissolution, *J. Biomed. Mater. Res.* **55**, 151-157.
73. ten Dijke P, Fu J, Schaap P, Roelen BA (2003) Signal transduction of bone morphogenetic proteins in osteoblast differentiation, *J. Bone Joint Surg. A.* **85-A (Supl. 3)**, 34-38.
74. Oldham, J.B., Lu, L., Zhu, X. and Porter, B.D. (2000) Biological activity of rhBMP-2 released from PLGA microspheres, *J. Biomech. Eng.* **122**, 289-292.
75. Ruhe, P.Q., Hedberg, E.L., Padron, N.T., Spauwen, P.H.M., Jansen, J.A. and Mikos, A.G. (2003) rhBMP-2 release from injectable poly(DL-lactic-co-glycolic acid)/calcium-phosphate cement composites, *J Bone Joint Surg Am.* **85-A Supl. 3**, 75-81.
76. Mori, M., Isobe, M., Yamazaki, Y., Ishihara, K. and Nakabayashi, N. (2000) Restoration of segmental bone defect in rabbit radius by biodegradable capsules containing recombinant human bone morphogenetic protein-2, *J. Biomed. Mater. Res.* **50**, 191-198.
77. Yang, X.B., Whitaker, M.J., Clarke, N.M.P., Sebald, W., Howdle, S.M., Shakesheff, K.M. and Oreffo, R.O.C. (2003) In vivo human bone and cartilage formation using porous polymer scaffolds encapsulated with bone morphogenetic protein-2, *J. Bone Miner. Res.* **18**, 1366.
78. Howdle, S.M., Watson, M.S., Whitaker, M.J., Popov, M.C., Davies, M.C., Mandel, F.S., Wang, J.D. and Shakesheff, K.M.(2001) Supercritical fluid mixing: preparation of thermally sensitive polymer composites containing bioactive materials, *Chem. Commun.* 109-110
79. Watson, M.S., Whitaker, M.J. Howdle, S.M. and Shakesheff, K.M. (2002) Incorporation of proteins into polymer materials by a novel supercritical fluid processing method, *Adv. Mater.* **14**, 1802-1804.
80. Effah-Kaufmann, E.A., Ducheyne, P. and Shapiro, I.M. (2000) Evaluation of osteoblast response to porous bioactive glass (45S5) substrates by RT-PCR analysis, *Tiss. Eng.* **6**, 19-28.
81. Sepulveda, P., Jones, J.R. and Hench, L.L. (2002) Bioactive sol-gel foams for tissue repair, *J. Biomed. Mater. Res.* **59**, 340-348.
82. Lee, K.Y., Peters, M.C., Anderson, K.W. and Mooney, D.J. (2000) Controlled growth factor release from synthetic extracellular matrices. *Nature* **408**, 998-1000.
83. Lee, K.Y. Peters, M.C. and Mooney, D.J. (2003) Comparison of vascular endothelial growth factor and basic fibroblast growth factor on angiogenesis in SCID mice. *J. Control. Release* **87**, 49-56.
84. Zisch, A.H., Lutolf, M.P., Ehrbar, M., Raeber, G.P., Rizzi, S.C., Davies, N., Schmokel, H., Bezuidenhout, D., Djonov, V., Zilla, P. and Hubbell, J.A. (2003) Cell-demanded release of VEGF from synthetic, biointeractive cell-ingrowth matrices for vascularized tissue growth, *FASEB J. Express 10.1096/fj.02-1041fje*, published online October 16, 2003.
85. Rangappa, N., Romereo, A., Nelson, K.D., Eberhart, R.C. and Smith, G.M. (2000) Laminin-coated poly(L-lactide) filaments induce robust neurite growth while providing directional orientation, *J. Biomed. Mater. Res.* **51**, 625-634.
86. Novikova, L.N., Novikov, L.N. and Kellerth, J.-O. (2003) Biopolymers and biodegradable smart implants for tissue regeneration after spinal cord injury, *Curr. Opin. Neurol.* **16**, 711-715.
87. Vunjak-Novakovic, G. (2003) The fundamentals of tissue engineering: scaffolds and bioreactors, *Novartis. Found. Symp.* **249**, 34-46.
88. Hayflick, L., Moorhead, P.S. (1961) The serial cultivation of human diploid strains, *Exp. Cell Res.* **25**, 585-621.
89. Moyzis, R.K., Buckingham, J.M., Cram, L.S., Dani, M., Deaven, L.L., Jones, M.D., Meyne, J., Ratliff, R.L. and Wu, J.R. (1988) A highly preserved repetitive DNA sequence (TTAGGG)$_n$, present at the telomeres of human chromosomes, *Proc. Natl. Acad. USA* **85**, 6622-6626.

90.   Hastie, N.D., Dempster, M., Dunlop, M.G., Thompson, A.M. Green, D.K. and Allshire, R.C. (1990) Telomere reduction in human colorectal carcinoma and with ageing, *Nature* **346**, 866-868.

91.   Harley, C.B. Futcher, A.B. and Gredier, C.W. (1990) Telomeres shorten during ageing of human fibroblasts, *Nature* **345**, 458-460.

92.   Bodnar, A.G., Ouellete, M., Frolkis, M., Holt, S.E., Chiu, C.P., Morin, G.B., Harley, C.B., Shay, J.W., Lichtsteiner, S. and Wright, W.E. (1998) Extension of life-span by introduction of telomerase into normal human cells, *Science* **279**, 349-353.

93.   Vaziri, H. and Benchimol, S. (1998) Reconstitution of telomerase activity in normal human cells leads to elongation of telomeres and extended replicative lifespan, *Curr. Biol.* **8**, 279-282.

94.   Jakob, M., Demarteau, O., Schafer, D., Stumm, M., Henerer, M. and Martin, L. (2003) Enzymatic digestion of adult human articular cartilage yields a small fraction of the total available cells, *Connect. Tissue Res.* **44**, 173-180.

95.   Pei M, Seidel J, Vunjak-Novakovic G, Freed LE (2002) Growth factors for sequential cellular de- and re-differentiation in tissue engineering, *Biochem. Biophys. Res. Commun.* **294**, 149-154.

96.   Lee DA, Reisler T, Bader DL (2003) Expansion of chondrocytes for tissue engineering in alginate beads enhances chondrocytic phenotype compared to conventional monolayer techniques, *Acta. Orthop. Scand.* **74**, 6-15.

97.   Wong M, Siegrist M, Goodwin K (2003) Cyclic tensile strain and cyclic hydrostatic pressure differentially regulate expression of hypertrophic markers in primary chondrocytes, *Bone* **33**, 685-693.

98.   Grimshaw, M.J. and Mason, R.M. (2001) Modulation of bovine articular chondrocyte gene expression in vitro by oxygen tension, *Osteoarthritis Cartilage* **9**, 357-364.

99.   Domm, C., Schunke, M., Christesen, K., and Kurz, B. (2002) Redifferentiation of dedifferentiated bovine articular chondrocytes in alginate culture under low oxygen tension, *Osteoarthritis Cartilage* **10**, 13-22.

100.   Bard, D.R., Dickens, M.J. Smith, A.U. and Zarek, J.M. (1972) Isolation of living cells from mature mammalian bone, *Nature* **236**, 314-315.

101.   Binderman, I., Duksin, D., Harell, A., Katzir, E. and Sachs, L. (1974) Formation of bone tissue in culture from isolated bone cells, *J Cell Biol.* **61**, 427-39.

102.   Chen, T.L., Cone, C.M. and Feldman D. (1983) Glucocorticoid modulation of cell proliferation in cultured osteoblast-like bone cells: differences between rat and mouse. *Endocrinology* **112**, 1739-1745.

103.   Chen, T.L., Cone, C.M. and Feldman, D. (1983) Effects of 1 alpha,25-dihydroxyvitamin D3 and glucocorticoids on the growth of rat and mouse osteoblast-like bone cells, *Calcif. Tissue. Int.*, **35**, 806-811.

104.   Cheng, H., Jiang, W., Phillips, F.M., Haydon, R.C., Peng, Y., Zhou, L., Luu, H.H., An, N., Breyer, B., Vanichakarn, P., Szatkowski, J.P., Park, J.Y. and He, T.C. (2003) Osteogenic activity of the fourteen types of human bone morphogenetic proteins (BMPs) *J. Bone Joint Surg.* **85-A**, 1544-1552.

105.   Stein, G.S., Lian, J.B., Stein, J.L., Van Wijnen, A.J. and Montecino, M. Transcriptional control of osteoblast growth and differentiation. *Physiol. Rev.* **76**, 593-629.

106.   Lian, J.B., Stein, G.S., Stein, J.L. and van Wijnen, A.J. Osteocalcin gene promoter: unlocking the secrets for regulation of osteoblast growth and differentiation. *J. Cell Biochem. Suppl.* **1998**, 62-72.

107.   Xu, C., Inokuma, M.S., Denham, J., Golds, K., Kundu, P., Gold, J.D. and Carpenter, M.K. (2001) Feeder-free growth of undifferentiated human embryonic stem cells, *Nat. Biotechnol.* **19**, 971-974.

108.   Tan, A., Romanska, H.M., Lenza, R., Jones, J., Hench, L.L., Polak, J.M. and Bishop, A.E. (2003) The effect of 58S bioactive sol-gel derived foams on the growth of murine lung epithelial cells, *Key Eng. Mater.* **240-2**, 719-723.

109.   Kim, H., Kim, H.W. and Suh, H. (2003) Sustained release of ascorbate-2-phosphate and dexamethasone from porous PLGA scaffolds for bone tissue engineering using mesenchymal stem cells, *Biomaterials* **24**, 4671-4679.

110.   Pryce, R.S. and Hench, L.L. (2003) Characterisation of a novel nitric oxide releasing bioactive glass, *Key. Eng. Mater.* **240-2**, 205-208.

111.   Cai, L., Okumu, F.W., Cleland, J.L., Beresini, M., Hogue, D., Lin, Z. and Filvaroff, E.H. (2002) A slow release formulation of insulin as a treatment for osteoarthritis, *Osteoarthritis Cartilage* **10**, 692-706.

112.   Yang, X.B., Roach, H.I., Clarke, N.M., Howdle, S.M., Quirk, R., Shakesheff, K.M. and Oreffo, R.O. (2001) Human osteoprogenitor growth and differentiation on synthetic biodegradable structures after surface modification, *Bone* **29**, 523-531.

113.   Levenberg, S., Huang, N.F., Lavik, E., Rogers, A.B., Itskovitz-Eldor, J. and Langer, R. (2003) Differentiation of human embryonic stem cells on three-dimensional polymer scaffolds, *Proc. Natl. Acad. Sci. USA* **100**, 12741-12746.

EMBRYONIC STEM CELLS FOR THE ENGINEERING AND REGENERATION OF MINERALIZED TISSUES

LEE D.K. BUTTERY & JULIA M. POLAK
Tissue Engineering and Regenerative Medicine Centre, ImperialCollege, Chelsea & Westminster Hospital, London, SWI0 9NH, UK

Key words: Embryonic stem cells, tissue engineering

1. Introduction

1.1. CLINICAL NEED/APPLICATIONS

Musculoskeletal diseases such as osteoporosis and arthritis or traumatic injuries can result in extensive tissue damage and are a major cause of morbidity. For example, there are an estimated 40,000 hip replacements performed in the United Kingdom each year and worldwide this is expected to exceed 6 million by 2050. Many fractures are associated with osteoporosis with one in four women sustaining at least one fracture by the age of 70. Collectively, diseases and traumas affecting the skeleton can severely restrict the quality of life and incur considerable healthcare costs. The need to address the increasing healthcare burden posed by musculoskeletal disorders culminated in the instigation of the Bone and Joint Decade 2000-2010, co-sponsored by the World Health Organization. A key goal of the initiative is advancing the understanding of musculoskeletal disorders through research to improve prevention and treatment. Tissue engineering is one area of research that is currently attracting considerable interest in the treatment of the damaged skeleton. In particular, recent advances in stem cell biology raises the prospect for generating unlimited numbers of specific cell types such as the osteoblast to enhance bone tissue repair and regeneration.

1.2. STEM CELLS

Stem cells are commonly defined as undifferentiated cells that can proliferate and have the capacity of both self-renewal and differentiation to one or more types of specialized cells. Recently, however the observation of de-differentiation and transdifferentiation of certain mature cells has led to re-evaluation of this definition [1-3].

For tissue engineering, stem cells provide a theoretically inexhaustible supply of cells

R.L. Reis and S. Weiner (eds.),
Learning from Nature How to Design New Implantable Biomaterials, 199-204.

that, depending on type, can give rise to some or all body tissues. Current research centres on promoting stem cell differentiation to the required lineage, derivation of highly purified cell populations, confirmation of a complete lack of carcinogenic potential and implantation in a form that will replace, or augment the function of, diseased or degenerating tissues [4-6]. The first step is selection of the most appropriate stem cell to form the required tissue. It is known that everyone carries around their own repository of stem cells that exist in various tissue niches, including bone marrow, brain, liver and skin as well as in the circulation [7-13]. Originally, these cells, sometimes referred to paradoxically as 'adult' stem cells, were considered to have only oligolineage potential but it is now known that they can show a considerable degree of plasticity [14,15] Thus, in theory, these cells could be harvested from a patient, incorporated into a tissue construct and put back in to the same individual when repair becomes necessary, bypassing the need for immunosuppression. Clearly, adult-derived progenitor cells need to be investigated and their clinical usefulness established. However, for some stem cell types, problems with accessibility, low frequency (e.g. in bone marrow there is roughly 1 stem cell per 100,000 cells), restricted differentiation potential and poor growth may limit their applicability to tissue engineering [6].

1.3. EMBRYONIC STEM CELLS

ES cells are harvested from the inner cell mass of the pre-implantation blastocyst and have been derived from rodents, primates and human beings [16-23]. Murine ES cells remain undifferentiated when grown in the presence of leukemia inhibitory factor (LIF) and, for some lines, culture on murine embryonic fibroblasts (MEF) [24,25]. LIF does not have the same effect on human ES cells and, in order to maintain them in an undifferentiated state, these require culture on MEF feeder layers in the presence of basic fibroblast growth factor (bFGF) or on Matrigel or laminin in MEF- conditioned medium [18,26]. When LIF or feeder cells are withdrawn, most types of ES cells differentiate spontaneously to form aggregates known, in view of their similarity to post-implantation embryonic tissues, as embryoid bodies. These spherical structures are comprised of derivatives of all three germ layers [27-30]. Synchronous formation of embryoid bodies, a prerequisite for gene expression studies of differentiating ES cells, can be achieved by removal of feeder cell layer or LIF followed by suspension culture.

1.4. MURINE ES CELLS

1.4.1. *Osteogenic Differentiation Potential*
Continued *in vitro* culture of murine ES cells as embryoid bodies leads to the formation of a range of differentiated cell types including cardiomyocytes [31] haematopoietic cells [32], neural cells [33], skeletal muscle [34], chondrocytes [35], adipocytes [36] and pancreatic islets [37]. In each case, despite the use of, for example, growth factors favouring the differentiation of a particular cell type, the resulting cultures were heterogeneous. Our own studies [38] have shown it is possible to derive cultures from murine ES cells that are enriched for osteoblasts. Initially, osteoblasts were generated from embryoid bodies, which were dispersed 5 days after removal of LIF, and grown in a culture medium designed for the maintenance and growth of explanted osteoblasts. Further experimentation revealed that not only does the biochemical composition of the

medium promote the differentiation to an osteoblast phenotype, but also that the time at which a particular stimulus is administered can enhance the effect. Thus, by providing the cells with dexamethasone 14 days following dispersal of embryoid bodies, it was possible to increase the yield of osteoblasts seven-fold [38].

In more recent studies we have found that following enrichment with osteoblastic cells by osteogenic stimuli it was is possible to isolate those cells using magnetic activated cell sorting methods with an antibody to cadherin11 [39]. Cadherin-11 is a cell adhesion molecule that is expressed by osteoblastic cells particularly in the early stages of differentiation. Subculture of the cadherin-ll positive cell population in the presence of osteogenic stimuli revealed that virtually all cells differentiated to form mature osteoblasts.

An alternative and potentially more efficient method of directing ES cell differentiation is genetic manipulation of the cells to overexpress a lineage specific transcription factor. This approach has been used to generate a range of cell types including cardiomyocytes, which formed stable grafts when implanted into mice with damaged hearts [40] Similarly, we have shown that overexpression of osterix (Tai, Polak, Chrostodoulou and Buttery, unpublished), an osteoblast transcription factor, can drive differentiation of ES cells into osteoblastic cells. Moreover, we have found that osterix can work synergistically with soluble factors like dexamethasone to further enhance differentiation. Taken together these data demonstrate that osterix transduction is useful not only for generating osteoblastic cells from ES cells but also investigating factors that influence this process and potentially delineating the ontogeny of the osteoblast and other mesodermal lineages.

1.5. HUMAN ES CELLS

1.5.1. *Osteogenic differentiation potential*
In recent studies we (Bielby, Polak and Buttery, unpublished) and others [41] have found that the treatment regime previously identified using mouse ES cells and in particular) the addition of dexamethasone at specific time points (Bielby, Polak and Buttery, unpublished) also induced robust osteogenic response from human ES cells *in vitro* (see fig 1). We identified mineralising cells *in vitro* and demonstrated the capacity of the cells, when implanted into SCID mice on a poly D,L lactide polymer scaffold, to give rise to mineralised tissue *in vivo* (Bielby, Polak and Buttery, unpublished; see fig 1). This data therefore demonstrates the derivation of osteoblast lineage cells from pluripotent human ES cells with the capacity to form mineralised tissue both *in vitro* and *in vivo*. Human ES cells can therefore differentiate *in vitro* in a way similar to mouse ES cells, and it should be feasible to further develop these methodologies to generate sufficient yields of osteogenic cells for use in skeletal tissue repair.

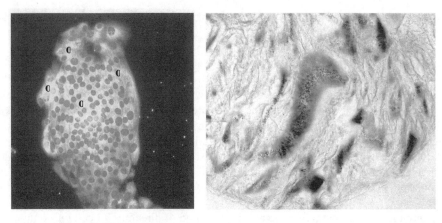

Figure 1. Osteogenic differentiation of human ES cells. Left panel shows formation of mineralized bone nodules *in vitro*, stained with osteocalcin (**o**) and DAPI nuclear counterstain (dots). Right panel shows formation of mineralized tissue in vivo after subcutaneous implantation of human ES cells into SCID mice on poly D,L, lactide sponge.

2. Future Perspectives

Our studies have demonstrated that osteoblasts and bone tissue can be generated from ES cells *in vitro* and *in vivo*. As such, ES are a useful cell source for investigating mechanisms of osteoblast differentiation and in the longer term might have applications the in repair of damaged bones. A number of studies are in progress to improve efficiency of osteogenic differentiation and in particular selection of proliferating osteoprogenitors and to be able to control this process under defined culture conditions. We are also looking at methods of controlling this process on three-dimensional scaffolds with a view to developing more accurate homologues (compared to monolayer culture) of endogenous bone tissue.

References

1. Bjornson, C.B., Rietze, R.L., Reynolds, B.A., Magli, M.C., Vecovi, A.I. (1999) Turning brain into blood: a haematopoietic fate adopted by adult neural stem cells in vivo, *Science* **283**, 534-537.
2. Mezey, E., Chandross, K.J., Harta, G., Maki, R.A., McKercher, S.R. (2000) Turning blood into brain: cells bearing neuronal antigens generated in vivo from bone marrow, . *Science* **290**, 1779-1782.
3. Toma, J., Akharvan, M., Fernandes, K.J.L., Barnabe-Heider, F., Sadikot, A., Kaplan, D.R., Miller, F.D. (2001) Isolation of multipotent adult stem cells from the dermis of mammalian skin, *Nature Cell Biol.* **3**, 778-784.
4. Blau, H.M., Brazelton, T.R., Weimann, J.M. (2001) The evolving concept of stem cells: entity or function?, *Cell* **105**, 829-841.
5. Odorico, J.S., Kaufman, D.S., Thomson, J.A. (2001) Multilineage differentiation from human embryonic stem cell lines, *Stem Cells* **19**, 193-204.
6. Bianco, P., Gehron, R.P. (2001) Stem cells in tissue engineering, *Nature* **414**, 118-121.
7. Lavker, R.M., Sun, T.T. (2000) Epidermal stem cells: properties, markers and location, *Proc. Nat.l Acad. Sci. USA* **97**, 13473-13475.
8. Uchida, N., Buck, D.W., He, D., Reitsma, M.J., Masek, M., Phan, T.V., Tsukamoto, A.S., Gage, F.H., Weissman, I.L. (2000) Direct isolation of human central nervous system stem cells, *Proc. Natl. Acad. Sci. USA* **97**, 14720-14725.
9. Vessey, C.J., de la Hall, P.M. (2001) Hepatic stem cells: a review, *Pathology* **33**, 130-141.
10. Weissman, I.L. (2000) Stem cells: units of development, units of regeneration and units in evolution,

Cell **100**, 157-168.

11. Lagasse, E., Shizuru, J.A., Uchida, N., Tsukamoto, A., Weissman, I.L. (2001) Toward regenerative medicine, *Immunity* **14**, 425-436.
12. McKay, R. (2000) Stem cells - hype and hope, *Nature* **406**, 361-364.
13. Van der Kooy, D., Weiss, S. (2000) Why stem cells?, *Science* **287**, 1439-1441.
14. Orkin, S.H. (2000) Stem cell Alchemy, *Nature Med.* **6**, 1212-1213.
15. Morrison, S.J. (2001) Stem cell potential: can anything, make anything?, *Curr. Biol. 11*, R7-9.
16. Martin, G. (1981) Isolation of a pluripotent cell line from early mouse embryos cultured in medium conditioned by teratocarcinoma cells, *Proc. Natl. Acad. Sci. USA* **78**, 7634-7638.
17. Evans, M.J., Kaufman, M.H. (1981) Establishment in culture of pluripotential stem cells from mouse embryos, *Nature* **291**, 154-156.
18. Thomson, I.A., Itskovitz-Eldor, J., Shapiro, S.S., Waknitz, M.A., Swiergiel, J.J., Marshall, V.S., Jones, M.J. (1998) Embryonic stem cell lines derived from human blastocysts, *Science* **282**, 1145-1147.
19. Reubinoff, B.E., Pera, M.F., Fong, C.Y., Trouson, A., Bongso, A. (2000) Embryonic stem cell lines ftom human blastocysts: somatic differentiation in vitro, *Nat. Biotechnol.* **18**, 399-404
20. Thomson, J.A., Kalishman, J., Golos, T.G., Durning, M., Harris, C.P., Becker, R.A., Hearn, J,P. (1995) Isolation of a primate embryonic stem cell line, *Proc. Natl. Acad. Sci. USA* **92**, 7844-7848.
21. Iannaconne, P.M., Taborn, G.U., Garton, R.L., Caplice, M.D., Brenin, D.R. (1994) Pluripotent embryonic stem cells from the rat are capable of producing chimeras, *Dev. Biol.* **163**, 288-292.
22. Graves, K.H., Moreadith, R.W. (1993) Derivation and charcaterization of putative pluripotential embryonic stem cells from pre-implantation rabbit embryos, *Mol. Reprod. Devel* **36**, 424-433.
23. Doetschman, T., Williams, P., Maeda, N. (1988) Establishment of hamster blastocyst- derived embryonic stem cells, *Dev. Biol.* **127**, 224-227.
24. Smith, A.G., Heath, J.K., Donaldson, D.D., Wong, G.G., Moreau, J., Stahl, M., Rogers, D. (1988) Inhibition of pluripotential embryonic stem cell differentiation by purified polypeptides, *Nature* **336**, 688-690.
25. Williams, R.L., Hilton, D.J., Pease, S., Wilson, T.A., Stewart, C.I., Gearing, D.P., Wagner, E.F., Metcafs, D., Nicola, N.A., Gough, N.M. (1988) Myeloid leukaemia inhibitory factor maintains the developmental potential of embryonic stem cells, *Nature* **336**, 684-687.
26. Xu, C., Inokumi, M.S., Denham, J., Golds, K., Kundu, P., Gold, J.D., Carpenter, M.K. (2001) Feeder-free growth of undifferentiated human embryonic stem cells, *Nature Biotech.* **19**, 971-974.
27. Doetschman, T.C., Eistetter, H., Katz, M., Schmidt, W., Kemler, R. (1985) The in vitro development of blastocyst-derived embryonic stem cell lines: formation of visceral yolk sac, blood islands and myocardium, *J. Embryol. Exp. Morphol.* **87**, 27-45.
28. Wartenberg, M., Gunther, J., Heschler, J., Sauer, H. (1998) The embryoid body as a novel in vitro assay system for anti-angiogenic agents, *Lab. Invest.* **78**, 1301-1314.
29. Desbaillets, I., Ziegler, U., Groscurth, P., Gassmann, M. (2000) Embryoid bodies: an in vitro model of mouse embryogenesis, *Exp. Physiol.* **85**, 645-651.
30. Itskovitz-Eldor, J., Schuldiner, M., Karsenti, D., Eden, A., Yanuka, O., Amit, M., Sorq, H., Benvenisty, N. (2000) Differentiation of human embryonic stem cells into embryoid bodies comprising the three embryonic germ layers, *Mol. Med.* **6**, 88-95.
31. Wobus, A.M., Wallukat, G., Heschler, J. (1991) Pluripotent mouse embryonic stem cells are able to differentiate into cardiomyocytes expressing chronotropic responses to adrenergic and cholinergic agents and Ca^{2+} channel blockers, *Differentiation* **48**, 73-182.
32. Keller, G., Kennedy, M., Papayannopoulou, T., Wiles, M.V. (1993) Haematopoeitic commitment during embryonic stem cell differentiation in culture, *Mol. Cell. Biol.* **13**, 473-486.
33. Bain, G., Kitchens, D., Yao, M., Huettner, J.E., Gottlieb, D.I. (1995) Embryonic stem cells express neuronal properties in vitro, *Devel Biol.* **168**, 342-357.
34. Rohwedel, J., Maltsev, V., Bober, E., Arnold, H.H., Hescheler, J., Wobus, A.M. (1994) Muscle cell differentiation of embryonic stem cells reflects myogenesis in vivo: developmentally regulated expression of myogenic determination genes and functional expression ionic currents, *Dev. Biol.* **164**, 87-101.
35. Kramer, J., Hegert, C., Guan, K., Wobus, A.M., Muller, P.K., Rohwedel, J. (2000) Embryonic stem cell-derived chondrogenic differentiation in vitro: activation by BMP-2 and BMP-4, *Mech. Dev.* **92**, 193-205.
36. Dani, C., Smith, A.G., Dessolin, S., Leroy, P., Staccini, L., Villageois, P., Darimont, C., Ailhard, G. (1997) Differentiation of embryonic stem cells into adipocytes in vitro, *J. Cell Sci.* **110**, 1279-1285.
37. Lumelsky, N., Blondel, O., Laeng, P., Velasco, I., Ravin, R., McKay, R. (2001) Ditferentiation of embryonic stem cells into insulin-secreting structures similar to pancreatic islets, *Science* **292**, 1389-1394.

38. Buttery, L.D.K., Boume, S., Xynos, J.D., Wood, H., Hughes, F.J., Hughes, S.P.F., Episkopou, V., Polak, J.M. (2001) Differentiation of osteoblasts and in vitro bone formation ITom murine embryonic stem cells, *Tiss. Eng.* **7**, 89-99.

39. Bourne, S., Polak, J.M., Hughes, S.P.F., Buttery, L.D.K. (2003) Osteogenic differentiation of mouse ES cells: Differential gene expression analysis by cDNA microarray and purification of osteoblasts by cadherin-11 magnetic activated cell sorting, Tiss Eng. in press

40. Klug, M.G., Soonpaa, M.H., Koh, G.Y., Field, L.J. (1996) Genetically selected cardiomyocytes from differentiating embryonic stem cells form stable intracardiac grafts, *J. Clin. Invest.* **98**, 216-224.

41. Sottile, V., Thomson, A., McWhir, J. (2003) Osteogenic differentiation of human ES cells, *Cloning Stem Cells* **5**, 149-155.

TISSUE ENGINEERING OF MINERALIZED TISSUES: THE ESSENTIAL ELEMENTS

A. J. SALGADO[1,2], M. E. GOMES[1,2,], R. L. REIS[1,2]

[1]3Bs Research Group - Biomaterials, Biodegradables and Biomimetics, University of Minho, 4710-057 Braga, Portugal
[2]Department of Polymer Engineering, University of Minho, 4800-058 Guimarães, Portugal

Abstract

Tissue Engineering (TE) has been emerging as a valid approach to the current therapies for hard tissue regeneration/substitution. In contrast to classic biomaterial approach, TE is based on the understanding of tissue formation and regeneration, and aims to induce new functional tissues, rather than just to implant new spare parts. The present chapter focuses on aspects that are believed to be essential for hard tissue engineering. Therefore, the use of cell transplantation and culturing on biodegradable scaffolds for the development of hybrid constructs aiming at the regeneration of hard tissues, like bone will be addressed. This review will also focus on the available in vitro systems for the culturing of cells-polymers constructs. Finally it will discussed the biofuctionality testing, and the several animal models that are available for evaluating the developed proofs of concept.

1. Introduction

As it was defined by Langer and Vacanti [1], tissue engineering is "an interdisciplinary field of research that applies the principles of engineering and the life sciences towards the development of biological substitutes that restore, maintain, or improve tissue function". In contrast to the classic biomaterials approach, it is based on the understanding of tissue formation and regeneration, and aims to induce new functional tissues, rather than just to implant new spare parts [2]. Researchers hope to reach this goal by combining knowledge from physics, chemistry, engineering, materials science, biology and medicine in an integrated manner [1-3].

One of the most widely studied hard tissue engineering approaches seekes to regenerate the lost or damaged tissue by making use of the interactions between cells and biodegradable scaffolds. This strategy usually involves the seeding and *in vitro* culturing of cells within a 3-D polymeric matrix – a scaffold - prior to implantation. The bioresorbable scaffold must be biocompatible and porous interconnected network to facilitate rapid vascularization and growth of newly formed tissue [3-8]. During the *in*

R.L. Reis and S. Weiner (eds.),
Learning from Nature How to Design New Implantable Biomaterials, 205-222.
© 2004 *Kluwer Academic Publishers. Printed in the Netherlands.*

vitro culture period, the seeded cells proliferate and secrete tissue specific extracellular matrix (ECM). Following implantation, the scaffold will gradually degrade and it will eventually be totally eliminated from the body [3-8].

2. Available Scaffolds

In scaffold-based tissue engineering strategies, the formation of new tissue is deeply influenced by the three dimensional environment provided by the scaffolds, namely its composition, porous architecture and, of course, its biological response to implanted cells and/or surrounding tissues. Therefore, the selection of an appropriate scaffold material is a major consideration [9].

Besides the obvious demands of biocompatibility and biodegradability, an ideal tissue engineering scaffold should have appropriate mechanical properties [4,10-14] and an adequate degradation rate [10,12-16]. Furthermore, the scaffold must possess adequate porosity, interconnectivity and permeability to allow the ingress of cells and nutrients [12,14-16] and the appropriate surface chemistry for enhanced cell attachment and proliferation [5,12,13,17]. For most applications, tissue engineering scaffolds must provide anchorage sites to cells, mechanical stability and structural guidance and, when implanted, provide an adequate interface to respond to physiological and biological changes in order to integrate with the surrounding native tissue.

In order to meet all the necessary requirements, scaffold materials must be fabricated from polymers with adequate properties. However, the establishment of basic requisites and design constrains its not an easy task and requires a deep knowledge about all the materials features that can interfere with cells/tissues-scaffold interactions. Many of these features are dictated by the processing methodology used to fabricate scaffolds. The processing of matrices, as well as the materials used in their development, to serve as templates for cell attachment/suspension and delivery has progressed at a tremendous rate in the past years and a wide range of methodologies have been developed [6,14,18-26]. Further details on this topic can be found elsewhere [27-29].

Several biodegradable polymers have been proposed to be used as three-dimensional scaffolds for bone tissue engineering, including a new range of natural origin polymers based on starch [20,21,30]. Starch-based polymers are degradable and biocompatible polymers [31-33] , with distinct structural forms and properties that can be tailored by the synthetic component of the starch-based blend, their processing methods, and the incorporation of additives and reinforcement materials. For this reason, together with their low cost and abundance of raw materials, starch-based polymers have been suggested for a wide range of biomedical applications. Several processing methodologies have been developed to produced starch based scaffolds with different properties and porous architectures, from conventional melt based technologies, such as extrusion and injection moulding using blowing agents [20,21] to inovative techniques, such as microwave baking [30]. Some of these scaffolds have been sucssufuly used in bone tissue engineering studies using human osteoblasts [34,35] and rat bone marrow stromal cells [36,37].

3. Bone Tissue Engineering Strategies

First, and even before starting the design and development of a scaffold, the researcher must chose an adequate tissue engineering strategy, which should be designed to mimic the natural process of repair. The following examples and discussion will be focusing on bone tissue engineering. Bone tissue engineering constructs should ultimately have two main functions when implanted *in vivo*: 1) provide structural support until the neotissue can assure it by itself and 2) promote osteoinduction, meaning in a simplistic way, the promotion of migration and differentiation of mesenchymal stem and osteprogenitor cells, which lately will lead to new bone formation [2,38,39]. In order to achieve such objectives three general approaches have been applied within the bone tissue engineering field [40]: 1)scaffold/matrix based therapies, 2)growth factors based therapies and 3)cell based therapies. Scaffold/matrix therapies are based on the implantation of structural implants on the injury site, relying on the endogenous recruitment of osteoprogenitor cells. This strategy does present, however, some problems, mainly due to the inefficacy of biodegradable polymers to act as osteoinductive agents [29,40]. It is known that ceramics, namely hydroylapatite (HA) and β-tricalcium phosphate (β-TCP) show osteoinductive potential [41-43], but it is also known that these may not fully possess other characteristics needed for bone tissue engineering applications, specially as a result of their brittle nature and inadequate degradation behaviour, as thoroughly discussed elsewhere [29]. In an attempt to overcome the lack of osteoinductive properties of the biodegradable scaffolds, new strategies based on encapsulation, binding and loading of 3D scaffolds with growth factors were developed [44-51]. Within the later, one of the most used options, is the use of bone morphogenetic proteins (BMPs) as delivery agent [46-48]. These proteins were firstly described by Urist and co-workers [51], and have previously shown to have osteoinductive behaviour, having a direct effect on the up regulation of Runx2, as recently reviewed by Franceschi and Xiao [52]. However, and although this approach is promising, it does present a problem [40]: the dose required to induce bone formation may vary from patient to patient, even under the same traumatic event, which may pose a problem for frequent clinical applications. As an alternative to both these methodologies, cell based therapies were studied. These methodologies are based on the seeding of cells with osteogenic potential on 3D scaffolds followed either by direct implantation on the injury site or by an *in vitro* culturing period upon which the cells/scaffold construct is implanted. Both of these methodologies have proved to be successful in animal models [36,53-59]. Among them, and in our opinion, the second is the one which may reveal a more promising future. This is due to the *in vitro* culture period, in which osteogenic cells will elaborate bone extracellular matrix on the surface of the scaffolds (Figure 1). As stated by Lind *et al*. [39] and Ogushi *et al*. [60], bone matrix is a storage medium for growth factors that actively participate in the activation and maintenance of cellular processes during bone formation. Furthermore, this type of construct would also have active osteogenic cells on its surface that could interact with other bone cell types, promoting in this sense a bone regenerative process very similar to the natural healing of bone. Being so, and by following the described strategy, it would also be possible to render an osteoinductive behaviour to a non-biodegradable polymer.

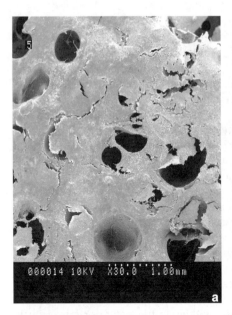

Figure 1. Rat bone marrow cells cultured on a starch/ethylene vinyl alcohol blend (SEVA-C) under osteogenic conditions over a period of 2 weeks: a) cells growing on the surface of the scaffold and b) cement line formation.

However, and in order to use this later strategy, an osteogenic cell population must be chosen. Theoretically speaking, osteoblasts, the "bone making cells", would be the chosen cell population. However, it should be herein noticed that the *in vitro* expansion potential of osteobasts is very low. Furthermore, in some bone related diseases the protein expression profile of osteoblasts may be irremediably affected, preventing their use for therapeutic purposes.

The non-hematopoetic fraction of bone marrow can be an alternative source of cells. It is known that within it, there are different cell populations with varying degrees of differentiation stages along the osteogenic lineage, including a population known as Mesenchymal Stem Cells (MSCs). MSCs have been several times proposed for bone tissue engineering applications [36,53-59]. It is believed that this cell population resides in the bone marrow, around blood vessels as pericytes [61,62]. One of the early problems regarding the study of MSC's, was the high heterogeneicity of whole bone marrow cultures. However, in 1992 Haynesworth *et al.* [63] described a method that, partially solved some of those. The methodology was based on the separation of the MSC's through gradient centrifugation. In later studies published by several authors [64-72] these cells were able to develop into distinct terminal and differentiated cells including bone [67-70], cartilage [66,67,70,71], fat [65,70] and tendon [64,72].

Besides its differentiation potential, MSC's present other important properties. As described by Bruder et al. [73] they can be extensively expanded *in vitro*. Later on Pittinger *et al.* [70] also showed that with an increased number of passages, they did not spontaneously differentiate. Furthermore it has been suggested that these cells may possess immunosuppressive effects which may render them either "immune privileged" or perhaps immunosuppressive roles *in vivo*, which would make them suitable for allogeneic or xenogeneic transplantation [74]. However, this subject still needs to be further investigated.

One of the problems related with the *in vitro* culture of these cells is that of their identification, other than using morphological evaluation. Recently several stem cell surface markers for the isolation and characterization of MSC were described. For instance, antibodies SB10, SH-2, SH-3 and SH-4 were found to bind to MSCs [75-78]. In 1999 Pittinger *et al.* described that human MSCs were shown to express a homogeneous (>98% purity) non-hematopoietic phenotype [70]. Furthermore they were also positive for SH-2, SH-3, CD71, CD44 and CD29 receptors [70]. Besides these markers, stem cells also express a myriad of cytokine, growth factors, extracellular matrix and adhesion related receptors, which makes difficult the establishment of universal markers for MSCs [79]. Further details on this subject can found in Table 1, and in the reviews by Ringe *et al.* [79] and by Roufosse *et al.* [80]. In a certain extent, this is due to the heterogeneity of the MSCs cultures, which possess different cell types with multilineage potential, even though to a different extent [79].

TABLE 1. Typical markers expressed by bone marrow mesenchymal stem cells. Adapted from references [79,80]

Marker Type	Notation
Specific Antigens	SH2, SH3, SH4, STRO-1, α smooth muscle actin, MAB1740
Extracellular Matrix	Collagen type I, III, IV, V, VI, proteoglycan, hyaluronan, fibronectin, laminin
Matrix Receptors	ALCAM, endoglin, hyaluronate receptor, ICAM-1, ICAM-2, VCAM-1, LFA-3, L-selectin
Adhesion Molecules	Integrins: $\alpha v\beta 3$, $\alpha v\beta 5$ Integrin chains: $\alpha 1$, $\alpha 2$, $\alpha 3$, $\alpha 4$, $\alpha 5$, $\beta 1$, $\beta 3$, $\beta 4$
Cytokines and Growth Factors	IL-1 α, -6, -7, -8, -11, -12, -13, -14, -15, LIF, SCF, Flt 3 ligand, GM-CSF, M-CSF
Cytokines and Growth Factors Receptors	IL-1R, -3R, -4R, -6R, -7R, PDGFR, TNFIR, TNFIIR, TGFβ1R, TGFβIIR, IFNγR, bFGFR, EGFR, LIFR, G-CSFR, SCFR, transferrin

Neverthless, and although they still present some problems, such as the low numbers upon isolation and the reduced differentiation capability of MSCs isolated from elderly patients, these have shown to have a vast potential to be used within the bone tissue engineering field, as previously reported by several authors [36,53-59].

4. *In Vitro* Culturing of Cells-Scaffold Constructs

Besides the selection of the scaffold material and the cell source (the two main components of this tissue engineering approach), it is necessary to develop more advanced procedures for growing cells in large quantities [6,81], optimizing the in vitro culturing systems currently used. The most widely used culturing technique in tissue engineering studies is static culturing which is often characterized by non-homogenous cell distribution, confining the majority of the cells to the outer surfaces of the scaffold, which in turn results to an inhomogeneous distribution of the *in vitro* generated extracellular matrix [82-84]. In order to overcome this limitation, several culturing systems which consist basically of growth chambers equipped with stirrers and sensors that regulate the appropriate amounts of nutrients, gases and waste products have been developed [82-84]. These systems, so-called bioreactors, may have different designs attempting to achieve one or more of the following objectives: i) maintain an uniform distribution of cells into the 3D scaffolds, ii) provide adequate levels of oxygen, nutrients, cytokines and growth factors iii) expose the cultured cells to mechanical stimuli. Furthermore, experiments involving in vitro bioreactor culturing can also be designed to study the effects of specific biochemical and physical signal involved in cell/tissue development and function, providing useful information on the processes that lead to the formation of 3-D tissues starting from cells/scaffolds tissue engineered constructs [82].

Bioreactors are also one of the focus of the development of a manufacturing technology for tissue engineered products, because they represent a chemically and mechanically controlled environment in which a tissue-like construct can be grown in reproducible conditions [81]. When the main purpose is to obtain engineered tissue-like substitutes, the type and the specific funtional design characteristics of a bioreactor are determined by the dimensional and functional requirements of the tissue to be substituted/regenerated as well as by the cell-scaffold system used.

There are several types of bioreactors currently available, which can be grouped in three main types, namely the spinner flasks, the rotating bioreators and the flow perfusion culture systems.

The **spinner flask** corresponds to one of the simplest biorreactor designs [84-88]. In these systems, the seeded scaffolds are attached to needles hanging from the cover of the flask and the mixing of the medium is maintained by a magnetic stir bar at the bottom of the flask. This mixing mechanism generates convective forces that enhance the nutrient concentration gradients but only at the surface of the scaffolds. Nevertheless, these systems have shown to increase the cell number on cartilage constucts based on chondrocytes and fibrous polyglycolic acid scaffolds, while under static culture conditions cell growth rates are diffusionally limited due to increasing cell mass and decreasing effective construct porosity resulting from cartilage matrix regeneration [89].

In another study [82], three-dimensional porous 75:25 poly(D,L-lactic-co-glycolic acid) biodegradable scaffolds were seeded with rat bone marrow cells (RBMCs) and cultured for 21 days under static conditions or in two model bioreactors (a spinner flask and a rotating wall vessel). The spinner flask culture demonstrated a 60% enhanced proliferation at the end of the first week when compared to static culture. Cell/polymer constructs cultured in the spinner flask had 2.4 times higher alkaline phosphatase (ALP) activity than constructs cultured under static conditions on day 14 and the total osteocalcin (OC) secretion in the spinner flask culture was 3.5 times higher than the static culture, with a peak OC secretion occurring on day 18. Furthermore, the spinner flask culture had the highest calcium content at day 14. On day 21, the calcium deposition in the spinner flask culture was 6.6 times higher than the static cultured constructs and over 30 times higher than the rotating wall vessel culture. Histological sections showed concentration of cells and mineralization at the exterior of the scaffolds at day 21. The accelerated proliferation and osteogenic differentiation of marrow cells, and the localization of the enhanced mineralization on the external surface of the scaffolds may be explained by the better mixing provided in the spinner flask, external to the outer surface of the scaffolds.

Spinner flasks can also be used as seeding sytems, generating more homogenous and controlled cell distribution and density on the scaffolds. This was observed, for example in a study where, highly porous, fibrous polyglycolic acid scaffolds, were seeded with bovine articular chondrocytes in spinner flasks. Essentially, all cells attached throughout the scaffold volume within 1 day. Mixing promoted the formation of 20-32-micron diameter cell aggregates that enhanced the kinetics of cell attachment without compromising the uniformity of cell distribution [87].

The **rotating wall vessel (RWV) bioreactor** was originally developed to protect delicate cell cultures from the high shear forces generated during the launch and landing of the space shuttle. Later on, when the device was tried for cell-line suspension cultures on the ground, cells were seen to aggregate and form larger structures ressembling tissues. This observation offered the exciting possibility that the bioreactor might be used to study the interactions of multiple cell types and their association with proliferation and cellular differentiation during the early steps of tissue formation [90].

Nowadays, the rotating wall vessel bioreactor can have several different designs and can be used with either microcarrier suspensions or scaffolds [91].

Basically, RWV bioreactor are horizontally rotated, fluid-filled culture vessels equipped with membrane diffusion gas exchange to optimize gas/oxygen supply. The initial rotational speed is adjusted so that the culture medium and the inoculum-individual cells, pre-aggregate cell constructs or tissue fragments-rotate synchronously with the vessel. As the cell aggregates grow in size, the rotational speed is increased to compensate for increased sedimentation rates. Under these conditions, at any given time, gravitational vectors are randomized and the shear stresses exerted by the fluid on the synchronously moving particles is minimized, thus establishing microgravity-like culturing conditions.

Two of the most well-known rotating wall vessel bioreactors designs are the High Aspect Ratio Vessel (HARV) and the Slow Lateral Turning Vessels (STLV) [92]. The HARV has a flat membrane oxygenator at the rear of the chamber and the SLVT consists of a cylindrical growth chamber containing an inner co-rotating cylindrical with a gas exchange membrane [92].
The oxygenation capacity of the HARV is higher than that of the SLTV and therefore, the HARV-type bioreactors are mostly used for cell types that require more oxygen per unit volume of culture medium while the SLTV-type bioreactors are suited for cells with low oxygen requirements. In a more advanced variant of the STLV, a fully automated computer controlled system continuously monitors flow through the rotating vessel, allowing for on-line monitoring of important parameters for cell development, such as pH, oxygen and glucose levels [92].

The high aspect ratio vessel (HARV) systems have been used to investigate the formation of 3-D rat marrow stromal cell culture on microcarriers, specifically bioactive ceramic hollow microspheres, under conditions of simulated microgravity [93-95] these systems are aimed at applications as microcarriers for bone tissue engineering and as drug delivery systems.

In one of these studies, hollow ceramic microspheres coated with synthesized calcium hydroxylapatite (HA) and sintered were developed and then placed in a rotating-wall vessel (RWV) bioreactor. The trajectory analysis revealed that the hollow microsphere remained suspended in the RWV bioreactor, and experienced a low shear stress (approximately 0.6 dyn/cm^2). The cell culture studies performed using rat bone marrow stromal cells and osteosarcoma cells (ROS 17/2.8) showed that the cells attached to and formed 3-D aggregates with the hollow microspheres under the culture conditions provided by the RWV bioreactor. In additon, extracellular matrix was observed in the aggregates [94].

In another study [94], it was investigated the formation of 3-D rat marrow stromal cell culture on microcarriers and the expression of bone-related biochemical markers under conditions of simulated microgravity, using a high aspect ratio vessel (HARV) system. In addition, it was calculated the shear stresses imparted on the surface of microcarriers of different densities by the medium fluid in a HARV. Again, the examination of cellular morphology by scanning electron microscopy revealed the presence of three-dimensional multicellular aggregates consisting of multiple cell-covered microcarriers bridged together. Mineralization was observed in the aggregates. The expressions of alkaline phosphatase activity, collagen type I, and osteopontin were shown via the use of histochemical staining, immunolabeling, and confocal scanning electron microscopy. Using a numerical approach, it was found that at a given rotational speed and for a given culture medium, a larger density difference between the microcarrier and the culture medium (e.g., a modified bioactive glass particle) imparted a higher maximum shear stress on the microcarrier.

The **flow perfusion bioreactor** is a bioreactor design that improves mass transfer at the interior of scaffolds [85]. The flow perfusion bioreactor uses a pump to perfuse medium continuously through the interconnected porous network of the seeded scaffold. The fluid path must be confined, so as to ensure the flow path is through the scaffold, rather than around the edges. The bioreactors that employ the latter flow path, i.e., exchanging medium in the chamber around the scaffold do not guarantee the exchange of medium within the interior of the scaffold and are termed "perfusion chambers" [86]. The perfusion bioreactor enhances the transport of nutrients because it allows the transport of medium through the interconnected pores of the scaffold. In additon, it offers a convenient way of providing mechanical stimulation to cells by means of fluid shear stress, which is particularly important in bone tissue engineering since bone cells are known to be stimulated by mechanical signals [96,97]. Furthermore, the magnitude of the shear stresses experienced by the cells can be varied by adjusting the flow rates throught the systems. However, the local shear stresses experienced by individual cells are also dependent on the scaffold microarchitecture [85].

Due to its characteristics, the flow perfusion bioreactor may facilitate the in-vitro development of tissue-like constructs for the regeneration of larger tissue defects. In additon, this culturing system also provides a valuable tool for *in vitro* investigation on biological mechanisms associated to bone growth and regeneration. In fact, the true biological environment of a bone cell derives from a dynamic interaction between responsively active cells experiencing mechanical forces and a continouslly changing 3D matrix architecture, which can be simulated, obviouslly to a limited extent, in this type of bioreactor.

Several studies [97-99] have been carried out aiming at studing the differentiation and proliferation patterns of marrow stromal cells cultured in 3D titanium meshes under flow perfusion conditions. These studies demonstrated that under flow conditions (at different flow rates), mineralized matrix production was dramatically increased over statically cultured constructs with the total calcium content of the cultured scaffolds increasing with increasing flow rate. Flow perfusion induced *de novo* tissue modelling with the formation of pore-like structures in the scaffolds and enhanced the distribution of cells and matrix throughout the scaffolds. These results report on the long-term effects of fluid flow on primary differentiating osteoblasts and indicate that fluid flow has far-reaching effects on osteoblast differentiation and phenotypic expression in vitro [97]. Further studies, using the same type of bioreactor, investigated the direct involvement of fluid shear stresses on the osteoblastic differentiation of marrow stromal cells. For this purpose, rat bone marrow stromal cells were seeded in 3D porous titanium fiber mesh scaffolds and cultured for 16 days in a flow perfusion bioreactor with perfusing culture media of different viscosities while maintaining the fluid flow rate constant. This methodology allowed exposure of the cultured cells to increasing levels of mechanical stimulation, in the form of fluid shear stress, whereas chemotransport conditions for nutrient delivery and waste removal remained essentially constant. Under similar chemotransport for the cultured cells in the 3D porous scaffolds, increasing fluid shear forces led to increased mineral deposition, suggesting that the mechanical stimulation provided by fluid shear forces in 3D flow perfusion culture can indeed enhance the expression of the osteoblastic phenotype. Increased fluid shear forces also resulted in the generation of a better spatially distributed extracellular matrix inside the porosity of the 3D titanium fiber mesh scaffolds. The combined effect of fluid shear forces on the mineralized extracellular matrix production and distribution emphasizes the importance of mechanosensation on osteoblastic cell function in a 3D environment [98].

Other studies [37] were performed using flow perfusion bioreactor as culturing systems for rat bone marrow stromal cells seeded onto two novel biodegradable scaffolds exhibiting distinct porous structures aiming at evaluating the potential of this tissue engineering approach for the generation of osteoinductive bone tissue replacement constructs. Specifically, scaffolds based on SEVA-C (a blend of starch with ethylene vinyl alcohol) and SPCL (a blend of starch with polycaprolactone) were examined in static and flow perfusion cultures. SEVA-C scaffolds were formed using an extrusion process [21], while SPCL scaffolds were obtained by a fiber bonding process [37]. These scaffolds were seeded with marrow stromal cells and cultured in a flow perfusion bioreactor and in 6-well plates up to 15 days. The calcium content analysis revealed a significant enhancement of calcium deposition on both scaffold types cultured under flow perfusion. This observation was confirmed by Von Kossa stained sections and tetracycline fluorescence. Histological analysis and confocal images of the cultured scaffolds showed a much better distribution of cells within the SPCL scaffolds than within the SEVA-C scaffolds, under flow perfusion conditions. In the scaffolds cultured under static conditions, only a surface layer of cells was observed. These results demonstrate the ability of the flow perfusion bioreactor to enhance the osteogenic differentiation and the homogeneous distribution of marrow stromal cells within starch-based polymer scaffolds. They also indicate that scaffold architecture and especially pore interconnectivity affect the homogeneity of the formed tissue. Accordingly, starch-based porous scaffolds seeded with mesenchymal stem cells and cultured under flow perfusion constitute a promising approach for the generation of osteoinductive bone tissue replacement constructs [37].

5. Biofunctionality Testing: Animal Models

After making the essential and initial *in vitro* tests, the proof of concept of a certain tissue engineering strategy must be tested *in vivo*. For this purpose a series of animal models and tests should be developed. These, commonly start by firstly using small animal models, followed by feasibility studies in larger animals and efficacy studies in non-human primates [100]. In principle, this thorough pathway of pre-clinical evaluation will give more details on how the hybrid constructs interact with the host's body, and if its utilization is feasible for clinical applications.

5.1. ECTOPIC MODELS

Among the first approaches are ectopic animal models, which have the advantage of the low costs involved and of being fairly simple to execute. Within these, the subcutaneous models (Figure 2), are particular popular, and have shown to be a scientifically well-established technique for construct osteogenicity [101,102]. In this particular group of animal models, the subcutaneous nude-mouse model is often used because it allows the implantation of tissue engineered hybrid constructs of other species' cells, due to the immune-compromised status of these mice [100]. Still within the same group of animal models, Wistar rats can be used. Although in this model it is not possible to use other species' cells, as in the previous case, there will not be a rejection process if the cells used are from the same species, particularly if obtained from the same litter. Still within the same group of animal models, other ectopical sites such as the muscle, peritoneal cavity, or mesentery can be used [103]. Furthermore, these ectopic models do also allow assessing if the scaffolds allow for an adequate vascularization, and also whether fibrous encapsulation of the scaffold occurs, as shown in Figure 3.

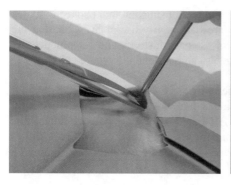
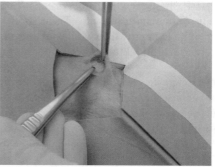

Figure 2. Representative example of a subcutaneous implantation procedure. (Images were a kind gift from Miss Wand Opreha).

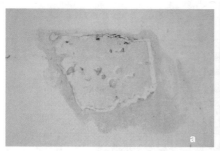

Figure 3. Histological micrographs (H&E staining) of a starch/cellulose (SCA) scaffold after 8 weeks of implantation in a subcutaneous model. Scaffolds were previously seeded with rat bone marrow cells. No fibrous encapsulation or any other kind of inflammatory response could detected(a) 8.6x and b)102.4x).

5.2. ORTHOPIC MODELS

As it is above described, the ectopic models are an useful tool for an initial evaluation of the proof of concept. However, they do present some limitations, and do not always directly correlate with the data obtained from orthopic experiments [100]. Being so, it is obvious that after this initial assessment more advanced and elaborated animal models have to be tested and developed, and there is a general consensus that the final preclinical animal models in which the new methodology is tested should mimic the clinical situation as close as possible. Because of this it is strongly advised the use of orthopic animal models, after the first initial ectopical assessment.

For an initial assessment on the evaluation of scaffolds behaviour on bony environments the bone chambers [104] and rat femoral defect models [105] are particularly useful for an initial orthopic evaluation.

Bone chambers can be a very useful model for testing scaffolds, in a well controlled environment and with a high level of reproducibility [104]. They are commonly made out of titanium and are placed in a bony environment, allowing the study of the scaffolds behaviour under non-load bearing conditions. As recently described by Buma *et al.* [104], these growth chambers consist of an outer housing and inner space in which the bone tissue can grow into. These models are particularly useful in evaluating the degree of bone in growth as well as the levels of osteoconductivity and osteoincuctivity of a certain scaffold. Further details on this models can be found elsewhere [104].

Another interesting models for the above referred objectives is the rat distal femoral defect in rats (Figure 4). It consists in drilling a hole of about 2.3mm diameter by 3.0mm height in the distal area of the femur. Although is not a critical size defect it is very useful for the determination of the inflammatory response caused by a scaffold, as well as for assessing the osteoconductivity, osteinductivity and bone bonding capability of the later [105].

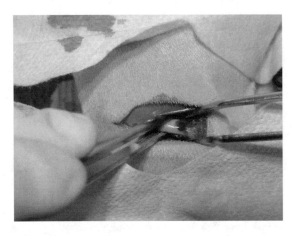

Figure 4. Non critical size defect on the rat distal area of the femur (Image was a kind gift from Dr. Vanessa Mendes).

If the object of study is the regeneration of cranio-facial lesions, the calvarian models, namely the rabit calvaria models, are often used [46,96]. A critical size defect (CSD) for this model is 15mm. This model is very popular and appropriate because it presents several advantages such as [103]: (1) the calvarial bone is a plate which allows the creation of a uniform circular defect that enables convenient radiographical and histological analysis; (2) the calvarial bone has a good size for easier surgical procedure and specimen handling; (3) no fixation is required because of the good supports by the dura and the overlying skin; (4) the model has been thoroughly used and studies, and is well reproduced; (5) it is relatively economical when compared to dogs and (6) as shown by Schantz *et al.* [54] up to two implants can be placed at the same time, allowing in this sense for accurate control samples under the same environment as the tested tissue engineering constructs.

Another particular useful model within this field, is the orbital defect model. For this purpose pigs are commonly the chosen animals, and it consists on the creation of a 2x2 cm defect on the orbital floor of the pigs, which as shown to be the critical size for this defects [106].

Large segmental bone defects are a very serious problem in clinical practice. In most cases associated to the bone loss there is an extensive degeneration of the surrounding periosteum, which as it is known is also a source of bone progenitor and stem cells. One of the most used animal models for segmental defects is the resection of a critical size part of the radius, presenting the following advantages [103]: (1) the radius bone is tubular, which allows the creation of segmental defects that enables convenient radiographical and histological analysis; (2) no fixation is required because of the support of the ulna and (3) is relatively economical. In this case a 15 mm defect is defined as CSD. However, it should be noticed that in this particular model the tissue engineering construct would not be under load-bearing conditions.

As an alternative to this, and if load bearing conditions want to be tested, the segmental defect in the rabbits femur is a valid alternative. In this case the segmental defect as the same length as the radius model (15mm) [53]. However, and because it is a load bearing area the implantation of the TE construct will need the adequate fixation, so when the animal starts to move the implant is not displaced due to the applied load. In order to overcome this problem titanium plates are used to stabilize the defect area. However, it should be reminded that their use may somewhat limit the range of post-dissection analytical tools, such as the traditional histological techniques.

Still within this class of defects, Pigs or sheeps models can and shoud also be used. However its utilization is rare, mainly due to the high costs implied to these models.

5.3. EVALUATION OF BONE REPAIR

Histological staining methodologies are the common method to assess bone formation. Common histological parameters include the following categories: bone union at the two osteotomies, callus formation, new bone formation of the defect, resorption of the bone graft, marrow changes and cortex remodelling. Histomorphometric analysis is also another technique that can be used to assess bone formation. By using computerized image analysis it is possible to quantify the area and penetration of bone tissue, are and thickness of non-mineralized bone like tissue, area of osteoblast covered surfaces, thickness of trabeculae, area and thickness of cartilage tissue, area of fibrovascular tissue and void space. Finally, radiographic analysis is also very useful to assess bone regeneration, namely if a follow up procedure is desired. With it, it is possible to obtain information on the amount and quality of the new bone, such as bone density and structure, and continuity with the adjacent recipient bone, can be obtained [103].

6. Final Remarks

Tissue engineering of bone and cartilage is in fact a promising field to overcome the short comes of the existing therapies of hard tissue replacement/regeneration. However, and so this objective can be successfully achieved, a scaffold with ideal structure and degradation properties has to be developed. Besides this it is also important to choose adequate cell populations, as well as, adequate conditions to culture the referred cell population within the scaffold material. Finally, it is necessary to develop and use adequate animal models that do not only allow the researcher to verify the developed concepts, but at the same time exactly mimic (as much as possible) the situations that are pretended to be regenerated. In sum, besides a profound knowledge of all the referred aspects, so that tissue engineering can be in a near future fact a solution, and not only a running promise, for the existing problems of bone and cartilage replacement/regeneration.

Acknowledgments

The authors would like to acknowledge Portuguese Foundation for Science and Technology through funds from POCTI and/or FEDER programs (PhD scholarship to A.J. Salgado, SFRH/BD/3139/2000, and to M.E. Gomes, SFRH/4704/2001). The authors would also like to acknowledge Dr. Vanessa Mendes and Miss Wanda Oprea for kindly helping on obtaining images to the present publication.

References

1. Langer, R. and Vacanti, J.P. (1993) Tissue Engineering, *Science* **260**, 920-926.
2. Kneser, U., Schaefer, D.J., Munder, B., Klemt, C., Andree, C. and Stark, G.B. (2002) Tissue engineering of bone, *Minimally Invasive Therapy & Allied Technologies* **11**, 107-116.
3. Laurencin, C.T., Ambrosio, A.M.A., Borden, M.D. and Cooper, J.A. (1999) Tissue engineering: Orthopedic applications, *Annual Review of Biomedical Engineering* **1**, 19-46.
4. Freed, L.E. and Vunjak-Novakovic, G. (1998) Culture of organized cell communities, *Advanced Drug Delivery Reviews* **33**, 15-30.
5. Langer, R. (1999) Selected advances in drug delivery and tissue engineering, *J Control Release* **62**, 7-11.
6. Lu, L.C. and Mikos, A.G. (1996) The importance of new processing techniques in tissue engineering, *Mrs Bulletin* **21**, 28-32.
7. Langer, R.S. and Vacanti, J.P. (1999) Tissue engineering: the challenges ahead, *Scientific American* **280**, 86-9.
8. Mooney, D.J. and Mikos, A.G. (1999) Growing new organs, *Scientific American* **280**, 38-43.
9. Hutmacher, D.W., Teoh, S.H., Zein, I., Ranawake, M. and Lau, S. (2000) Tissue engineering research: the engineer's role, *Med Device Technol* **11**, 33-9.
10. Middleton, J.C. and Tipton, A.J. (2000) Synthetic biodegradable polymers as orthopedic devices, *Biomaterials* **21**, 2335-46.
11. Vacanti, C.A. and Bonassar, L.J. (1999) An overview of tissue engineered bone, *Clin Orthop* S375-81.
12. Kim, B.S. and Mooney, D.J. (1998) Development of biocompatible synthetic extracellular matrices for tissue engineering, *Trends in Biotechnology* **16**, 224-230.
13. Chapekar, M. (2000) Tissue engineering: challenges and opportunities, *Journal of Biomedical Materials Research (Applied Biomaterials)* **53**, 617-620.
14. Thomson, R.C., Yaszemski, M.J., Powers, J.M. and Mikos, A.G. (1995) Fabrication of Biodegradable Polymer Scaffolds to Engineer Trabecular Bone, *Journal of Biomaterials Science-Polymer Edition* **7**, 23-38.
15. Thompson, R., Wake, M.C., Yaszemski, M. and Mikos, A.G. (1995) Biodegradable polymer scaffolds to regenerate organs, *Advanced Polymer Science* **122**, 247-274.
16. Hutmacher, D.W. (2000) Scaffolds in tissue engineering bone and cartilage, *Biomaterials.* **21(24)**, 2529-43.
17. Freed, L.E., Vunjak-Novakovic, G., Biron, R.J., Eagles, D.B., Lesnoy, D.C., Barlow, S.K. and Langer, R. (1994) Biodegradable polymer scaffolds for tissue engineering, *Biotechnology (N Y)* **12**, 689-93.
18. Hutmacher, D.W., Schantz, T., Zein, I., Ng, K.W., Teoh, S.H. and Tan, K.C. (2001) Mechanical properties and cell cultural response of polycaprolactone scaffolds designed and fabricated via fused deposition modeling, *Journal of Biomedical Materials Research* **55**, 203-216.
19. Mikos, A.G., Thorsen, A.J., Czerwonka, L.A., Bao, Y., Langer, R., Winslow, D.N. and Vacanti, J.P. (1994) Preparation and Characterization of Poly(L-Lactic Acid) Foams, *Polymer* **35**, 1068-1077.
20. Gomes, M.E., Ribeiro, A.S., Malafaya, P.B., Reis, R.L. and Cunha, A.M. (2001) A new approach based on injection moulding to produce biodegradable starch-based polymeric scaffolds: morphology, mechanical and degradation behaviour, *Biomaterials* **22**, 883-889.

21. Gomes, M.E., Godinho, J.S., Tchalamov, D., Cunha, A.M. and Reis, R.L. (2002) Alternative tissue engineering scaffolds based on starch: processing methodologies, morphology, degradation and mechanical properties, *Materials Science & Engineering C-Biomimetic and Supramolecular Systems* **20**, 19-26.

22. Mooney, D.J., Baldwin, D.F., Suh, N.P., Vacanti, L.P. and Langer, R. (1996) Novel approach to fabricate porous sponges of poly(D,L-lactic-co-glycolic acid) without the use of organic solvents, *Biomaterials* **17**, 1417-1422.

23. Malafaya, P.B. and Reis, R.L. (2003) Bioceramics 15, in (eds.), *Porous bioactive composites from marine origin based in chitosan and hydroxylapatite particles*, pp. 39-42.

24. Taboas, J.M., Maddox, R.D., Krebsbach, P.H. and Hollister, S.J. (2003) Indirect solid free form fabrication of local and global porous, biomimetic and composite 3D polymer-ceramic scaffolds, *Biomaterials* **24**, 181-194.

25. Landers, R. and Mulhaupt, R. (2000) Desktop manufacturing of complex objects, prototypes and biomedical scaffolds by means of computer-assisted design combined with computer-guided 3D plotting of polymers and reactive oligomers, *Macromolecular Materials and Engineering* **282**, 17-21.

26. Holy, C.E., Dang, S.M., Davies, J.E. and Shoichet, M.S. (1999) In vitro degradation of a novel poly(lactide-co-glycolide) 75/25 foam, *Biomaterials* **20**, 1177-1185.

27. Agrawal, C.M. and Ray, R.B. (2001) Biodegradable polymeric scaffolds for musculoskeletal tissue engineering, *Journal of Biomedical Materials Research* **55**, 141-150.

28. Yang, S.F., Leong, K.F., Du, Z.H. and Chua, C.K. (2001) The design of scaffolds for use in tissue engineering. Part 1. Traditional factors, *Tissue Engineering* **7**, 679-689.

29. Salgado, A.J., Coutinho, O.P. and Reis, R.L. (2004) Bone Tissue Engineering: State of the Art and Future Trends, *Macromolecular Bioscience* submitted.

30. Malafaya, P.B., Elvira, C., Gallardo, A., San Roman, J. and Reis, R.L. (2001) Porous starch-based drug delivery systems processed by a microwave route, *Journal of Biomaterials Science-Polymer Edition* **12**, 1227-1241.

31. Gomes, M.E., Reis, R.L., Cunha, A.M., Blitterswijk, C.A. and de Bruijn, J.D. (2001) Cytocompatibility and response of osteoblastic-like cells to starch-based polymers: effect of several additives and processing conditions, *Biomaterials* **22**, 1911-1917.

32. Marques, A.P., Reis, R.L. and Hunt, J.A. (2002) The biocompatibility of novel starch-based polymers and composites: in vitro studies, *Biomaterials* **23**, 1471-1478.

33. Mendes, S.C., Reis, R.L., Bovell, Y.P., Cunha, A.M., van Blitterswijk, C.A. and de Bruijn, J.D. (2001) Biocompatibility testing of novel starch-based materials with potential application in orthopaedic surgery: a preliminary study, *Biomaterials* **22**, 2057-2064.

34. Salgado, A.J., Figueiredo, J.E., Coutinho, O.P. and Reis, R.L. (2004) Biological Response to Pre-Mineralized Starch Based Scaffolds for Bone Tissue Engineering, *Journal of Materials Science: Materials in Medicine* accepted for publication.

35. Salgado, A.J., Gomes, M.E., Chou, A., Coutinho, O.P., Reis, R.L. and Hutmacher, D.W. (2002) Preliminary study on the adhesion and proliferation of human osteoblasts on starch-based scaffolds, *Materials Science & Engineering C-Biomimetic and Supramolecular Systems* **20**, 27-33.

36. Mendes, S.C., Bezemer, J., Claase, M.B., Grijpma, D.W., Bellia, G., Degli-Innocenti, F., Reis, R.L., De Groot, K., Van Blitterswijk, C.A. and De Bruijn, J.D. (2003) Evaluation of two biodegradable polymeric systems as substrates for bone tissue engineering, *Tissue Engineering* **9**, S91-S101.

37. Gomes, M.E., Sikavitsas, V.I., Behravesh, E., Reis, R.L. and Mikos, A.G. (2003) Effect of flow perfusion on the osteogenic differentiation of bone marrow stromal cells cultured on starch-based three-dimensional scaffolds, *Journal of Biomedical Materials Research Part A* **67A**, 87-95.

38. Albrektsson, T. and Johansson, C. (2001) Osteoinduction, osteoconduction and osseointegration, *European Spine Journal* **10**, S96-S101.

39. Lind, M. and Bunger, C. (2001) Factors stimulating bone formation, *European Spine Journal* **10**, S102-S109.

40. Bruder, S.P. and Fox, B.S. (1999) Tissue engineering of bone - Cell based strategies, *Clinical Orthopaedics and Related Research* S68-S83.

41. Gosain, A.F., Song, L.S., Riordan, P., Amarante, M.T., Nagy, P.G., Wilson, C.R., Toth, J.M. and Ricci, J.L. (2002) A 1-year study of osteoinduction in hydroxyapatite-derived Biomaterials in an adult sheep model: Part I, *Plastic and Reconstructive Surgery* **109**, 619-630.

42. Yuan, H.P., de Bruijn, J.D., Zhang, X.D., van Blitterswijk, C.A. and de Groot, K. (2001) Bone induction by porous glass ceramic made from Bioglass (R) (45S5), *Journal of Biomedical Materials Research* **58**, 270-276.

43. Yuan, H.P., Kurashina, K., de Bruijn, J.D., Li, Y.B., de Groot, K. and Zhang, X.D. (1999) A preliminary study on osteoinduction of two kinds of calcium phosphate ceramics, *Biomaterials* **20**, 1799-1806.

44. Zhang, J.Y., Doll, B.A., Beckman, E.J. and Hollinger, J.O. (2003) Three-dimensional biocompatible ascorbic acid-containing scaffold for bone tissue engineering, *Tissue Engineering* **9**, 1143-1157.
45. Burdick, J.A., Mason, M.N., Hinman, A.D., Thorne, K. and Anseth, K.S. (2002) Delivery of osteoinductive growth factors from degradable PEG hydrogels influences osteoblast differentiation and mineralization, *Journal of Controlled Release* **83**, 53-63.
46. Han, B., Perelman, N., Tang, B.W., Hall, F., Shors, E.C. and Nimni, M.E. (2002) Collagen-targeted BMP3 fusion proteins arrayed on collagen matrices or porous ceramics impregnated with Type I collagen enhance osteogenesis in a rat cranial defect model, *Journal of Orthopaedic Research* **20**, 747-755.
47. Whang, K., Tsai, D.C., Nam, E.K., Aitken, M., Sprague, S.M., Patel, P.K. and Healy, K.E. (1998) Ectopic bone formation via rhBMP-2 delivery from porous bioabsorbable polymer scaffolds, *Journal of Biomedical Materials Research* **42**, 491-499.
48. Saito, N. and Takaoka, K. (2003) New synthetic biodegradable polymers as BMP carriers for bone tissue engineering, *Biomaterials* **24**, 2287-2293.
49. Park, Y.J., Lee, Y.M., Park, S.N., Sheen, S.Y., Chung, C.P. and Lee, S.J. (2000) Platelet derived growth factor releasing chitosan sponge for periodontal bone regeneration, *Biomaterials* **21**, 153-159.
50. Ueda, H., Hong, L., Yamamoto, M., Shigeno, K., Inoue, M., Toba, T., Yoshitani, M., Nakamura, T., Tabata, Y. and Shimizu, Y. (2002) Use of collagen sponge incorporating transforming growth factor-beta 1 to promote bone repair in skull defects in rabbits, *Biomaterials* **23**, 1003-1010.
51. Urist, M.R., Delange, R.J. and Finerman, G.A.M. (1983) Bone Cell-Differentiation and Growth-Factors, *Science* **220**, 680-686.
52. Franceschi, R.T. and Xiao, G.Z. (2003) Regulation of the osteoblast-specific transcription factor, runx2: Responsiveness to multiple signal transduction pathways, *Journal of Cellular Biochemistry* **88**, 446-454.
53. Holy, C.E., Fialkov, J.A., Davies, J.E. and Shoichet, M.S. (2003) Use of a biomimetic strategy to engineer bone, *Journal of Biomedical Materials Research Part A* **65A**, 447-453.
54. Schantz, J.T., Hutmacher, D.W., Lam, C.X.F., Brinkmann, M., Wong, K.M., Lim, T.C., Chou, N., Guldberg, R.E. and Teoh, S.H. (2003) Repair of calvarial defects with customised tissue-engineered bone grafts - II. Evaluation of cellular efficiency and efficacy in vivo, *Tissue Engineering* **9**, S127-S139.
55. Quarto, R., Mastrogiacomo, M., Cancedda, R., Kutepov, S.M., Mukhachev, V., Lavroukov, A., Kon, E. and Marcacci, M. (2001) Repair of large bone defects with the use of autologous bone marrow stromal cells, *New England Journal of Medicine* **344**, 385-386.
56. IshaugRiley, S.L., Crane, G.M., Gurlek, A., Miller, M.J., Yasko, A.W., Yaszemski, M.J. and Mikos, A.G. (1997) Ectopic bone formation by marrow stromal osteoblast transplantation using poly(DL-lactic-co-glycolic acid) foams implanted into the rat mesentery, *Journal of Biomedical Materials Research* **36**, 1-8.
57. Kadiyala, S., Jaiswal, N. and Bruder, S.P. (1997) Culture-expanded, bone marrow-derived mesenchymal stem cells can regenerate a critical-sized segmental bone defect, *Tissue Engineering* **3**, 173-185.
58. Bruder, S.P., Kraus, K.H., Goldberg, V.M. and Kadiyala, S. (1998) The effect of implants loaded with autologous mesenchymal stem cells on the healing of canine segmental bone defects, *Journal of Bone and Joint Surgery-American Volume* **80A**, 985-996.
59. Breitbart, A.S., Grande, D.A., Kessler, R., Ryaby, J.T., Fitzsimmons, R.J. and Grant, R.T. (1998) Tissue engineered bone repair of calvarial defects using-cultured periosteal cells, *Plastic and Reconstructive Surgery* **101**, 567-574.
60. Ohgushi, H. and Caplan, A.I. (1999) Stem cell technology and bioceramics: From cell to gene engineering, *Journal of Biomedical Materials Research* **48**, 913-927.
61. Caplan, A.I. and Bruder, S.P. (2001) Mesenchymal stem cells: building blocks for molecular medicine in the 21st century, *Trends in Molecular Medicine* **7**, 259-264.
62. Short, B., Brouard, N., Occhiodoro-Scott, T., Ramakrishnan, A. and Simmons, P.J. (2003) Mesenchymal stem cells, *Archives of Medical Research* **34**, 565-571.
63. Haynesworth, S.E., Goshima, J., Goldberg, V.M. and Caplan, A.I. (1992) Characterization of Cells with Osteogenic Potential from Human Marrow, *Bone* **13**, 81-88.
64. Awad, H.A., Butler, D.L., Boivin, G.P., Smith, F.N.L., Malaviya, P., Huibregtse, B. and Caplan, A.I. (1999) Autologous mesenchymal stem cell-mediated repair of tendon, *Tissue Engineering* **5**, 267-277.
65. Endres, M., Hutmacher, D.W., Salgado, A.J., Kaps, C., Ringe, J., Reis, R.L., Sittinger, M., Brandwood, A. and Schantz, J.T. (2003) Osteogenic induction of human bone marrow-derived mesenchymal progenitor cells in novel synthetic polymer-hydrogel matrices, *Tissue Engineering* **9**, 689-702.
66. Mackay, A.M., Beck, S.C., Murphy, J.M., Barry, F.P., Chichester, C.O. and Pittenger, M.F. (1998) Chondrogenic differentiation of cultured human mesenchymal stem cells from marrow, *Tissue Engineering* **4**, 415-428.

67. Kadiyala, S., Young, R.G., Thiede, M.A. and Bruder, S.P. (1997) Culture expanded canine mesenchymal stem cells possess osteochondrogenic potential in vivo and in vitro, *Cell Transplantation* **6**, 125-134.
68. Nilsson, S.K., Dooner, M.S., Weier, H.U., Frenkel, B., Lian, J.B., Stein, G.S. and Quesenberry, P.J. (1999) Cells capable of bone production engraft from whole bone marrow transplants in nonablated mice, *Journal of Experimental Medicine* **189**, 729-734.
69. Jaiswal, N., Haynesworth, S.E., Caplan, A.I. and Bruder, S.P. (1997) Osteogenic differentiation of purified, culture-expanded human mesenchymal stem cells in vitro, *Journal of Cellular Biochemistry* **64**, 295-312.
70. Pittenger, M.F., Mackay, A.M., Beck, S.C., Jaiswal, R.K., Douglas, R., Mosca, J.D., Moorman, M.A., Simonetti, D.W., Craig, S. and Marshak, D.R. (1999) Multilineage potential of adult human mesenchymal stem cells, *Science* **284**, 143-147.
71. Johnstone, B., Hering, T.M., Caplan, A.I., Goldberg, V.M. and Yoo, J.U. (1998) In vitro chondrogenesis of bone marrow-derived mesenchymal progenitor cells, *Experimental Cell Research* **238**, 265-272.
72. Young, R.G., Butler, D.L., Weber, W., Caplan, A.I., Gordon, S.L. and Fink, D.J. (1998) Use of mesenchymal stem cells in a collagen matrix for Achilles tendon repair, *Journal of Orthopaedic Research* **16**, 406-413.
73. Bruder, S.P., Jaiswal, N. and Haynesworth, S.E. (1997) Growth kinetics, self-renewal, and the osteogenic potential of purified human mesenchymal stem cells during extensive subcultivation and following cryopreservation, *Journal of Cellular Biochemistry* **64**, 278-294.
74. Devine, S.M. (2002) Mesenchymal stem cells: Will they have a role in the clinic?, *Journal of Cellular Biochemistry* 73-79.
75. Haynesworth, S.E., Baber, M.A. and Caplan, A.I. (1992) Cell-Surface Antigens on Human Marrow-Derived Mesenchymal Cells Are Detected by Monoclonal-Antibodies, *Bone* **13**, 69-80.
76. Bruder, S.P., Horowitz, M.C., Mosca, J.D. and Haynesworth, S.E. (1997) Monoclonal antibodies reactive with human osteogenic cell surface antigens, *Bone* **21**, 225-235.
77. Bruder, S.P., Ricalton, N.S., Boynton, R.E., Connolly, T.J., Jaiswal, N., Zala, J. and Barry, F.P. (1998) Mesenchymal stem cell surface antigen SB-10 corresponds to activated leukocyte cell adhesion molecule and is involved in osteogenic differentiation, *Journal of Bone and Mineral Research* **13**, 655-663.
78. Barry, F., Boynton, R., Murphy, M., Haynesworth, S. and Zaia, J. (2002) The SH-3 and SH-4 antibodies recognize distinct epitopes on CD73 from human mesenchymal stem cells (vol 289, pg 519, 2001), *Biochemical and Biophysical Research Communications* **290**, 1609-1609.
79. Ringe, J., Kaps, C., Burmester, G.R. and Sittinger, M. (2002) Stem cells for regenerative medicine: advances in the engineering of tissues and organs, *Naturwissenschaften* **89**, 338-351.
80. Roufosse, C.A., Direkze, N.C. and Otto, W.R., Wright N.A. (2004) Circulating Mesenchymal Stem Cells, *The International Journal of Biochemistry & Cell Biology* in press.
81. Vacanti, J.P. and Vacanti, C.A. (1997) Principles of Tissue Engineering, in R. Lanza, R. Langer and W. Chick (eds.), Academic Press, New York, pp. 1.
82. Sikavitsas, V.I., Bancroft, G.N. and Mikos, A.G. (2002) Formation of three-dimensional cell/polymer constructs for bone tissue engineering in a spinner flask and a rotating wall vessel bioreactor, *Journal of biomedical materials research.* **62**, 136-48.
83. Botchwey, E.A., Pollack, S.R., Levine, E.M. and Laurencin, C.T. (2001) Bone tissue engineering in a rotating bioreactor using a microcarrier matrix system, *Journal of biomedical materials research.* **55(2)**, 242-53.
84. Goldstein, A.S., Juarez, T.M., Helmke, C.D., Gustin, M.C., Mikos, A.G., Department of Bioengineering, I.o.B., Bioengineering, R.U.H.T.X.U.S.A. and Rice U, H.T.X. (2001) Effect of convection on osteoblastic cell growth and function in biodegradable polymer foam scaffolds, *Biomaterials.* **22(11)**, 1279-88.
85. Bancroft, G.N., Sikavitsas, V.I. and Mikos, A.G. (2003) Design of a flow perfusion bioreactor system for bone tissue-engineering applications, *Tissue Engineering* **9**, 549.
86. Freed, L.E. and Vunjak-Novakovic, G. (2000) Principles of Tissue Engineering, in R. Lanza, R. Langer and J. Vacanti (eds.), *Tissue engineering bioreactors*, Academic Press, San Diego, pp. 143-156.
87. Vunjak-Novakovic, G., Obradovic, B., Martin, I., Bursac, P.M., Langer, R. and Freed, L.E. (1998) Dynamic cell seeding of polymer scaffolds for cartilage tissue engineering, *Biotechnology progress.* **14(2)**, 193-202.
88. Vunjak-Novakovic, G., Martin, I., Obradovic, B., Treppo, S., Grodzinsky, A.J., Langer, R. and Freed, L.E. (1999) Bioreactor cultivation conditions modulate the composition and mechanical properties of tissue-engineered cartilage, *Journal of Orthopaedic Research* **17**, 130-138.

89. Freed, L.E., Vunjak-Novakovic, G., Langer, R., Division of Health, S. and Technology, M.I.o.T.C. (1993) Cultivation of cell-polymer cartilage implants in bioreactors, *Journal of cellular biochemistry.* **51(3)**, 257-64.
90. Duray, P., Hatfill, S. and Pellis, N. (1997) Tissue culture in microgravity, *Science & Medicine* 46-55.
91. Freed, L.E., Langer, R., Martin, I., Pellis, N.R. and Vunjak-Novakovic, G. (1997) Tissue engineering of cartilage in space, *Proceedings of the National Academy of Sciences of the United States of America.* **94(25)**, 13885-90.
92. Unsworth, B.R. and Lelkes, P.I. (1998) Growing tissues in microgravity, *Nat Med* **4**, 901-7.
93. Qiu, Q., Ducheyne, P., Gao, H. and Ayyaswamy, P. (1998) Formation and differentiation of three-dimensional rat marrow stromal cell culture on microcarriers in a rotating-wall vessel, *Tissue Eng* **4**, 19-34.
94. Qiu, Q.Q., Ducheyne, P. and Ayyaswamy, P.S. (1999) Fabrication, characterization and evaluation of bioceramic hollow microspheres used as microcarriers for 3-D bone tissue formation in rotating bioreactors, *Biomaterials* **20**, 989-1001.
95. Qiu, Q.Q., Ducheyne, P. and Ayyaswamy, P.S. (2001) 3D bone tissue engineered with bioactive microspheres in simulated microgravity, *In Vitro Cell Dev Biol Anim* **37**, 157-65.
96. Sikavitsas, V.I., Temenoff, J.S. and Mikos, A.G. (2001) Biomaterials and bone mechanotransduction, *Biomaterials* **22**, 2581-2593.
97. Bancroft, G.N., Sikavitsas, V.I., van den Dolder, J., Sheffield, T.L., Ambrose, C.G., Jansen, J.A. and Mikos, A.G. (2002) Fluid flow increases mineralized matrix deposition in 3D perfusion culture of marrow stromal osteoblasts in a dose-dependent manner, *Proceedings of the National Academy of Sciences of the United States of America.* **99(20)**, 12600-5.
98. Sikavitsas, V.I., van den Dolder, J., Bancroft, G.N., Jansen, J.A. and Mikos, A.G. (2003) Influence of the in vitro culture period on the in vivo performance of cell/titanium bone tissue-engineered constructs using a rat cranial critical size defect model, *Journal of biomedical materials research.* **67A**, 944-51.
99. Sikavitsas, V.I., Bancroft, G.N., Holtorf, H.L., Jansen, J.A. and Mikos, A.G. (2003) Mineralized matrix deposition by marrow stromal osteoblasts in 3D perfusion culture increases with increasing fluid shear forces, *Proceedings of the National Academy of Sciences of the United States of America.* **100**, 14683-8.
100. Kruyt, M.C., van Gaalen, S.M., Oner, F.C., Verbout, A., de Bruijn, J.D. and Dhert, W.J.A. (2004) Bone tissue engineering and spinal fusion: the potential of hybrid constructs by combining osteoprogenitor cells and scaffolds, *Biomaterials* **25**, 1463-1473.
101. Krebsbach, P.H., Kuznetsov, S.A., Satomura, K., Emmons, R.V.B., Rowe, D.W. and Robey, P.G. (1997) Bone formation in vivo: Comparison of osteogenesis by transplanted mouse and human marrow stromal fibroblasts, *Transplantation* **63**, 1059-1069.
102. Allay, J.A., Dennis, J.E., Haynesworth, S.E., Majumdar, M.K., Clapp, D.W., Shultz, L.D., Caplan, A.I. and Gerson, S.L. (1997) LacZ and interleukin-3 expression in vivo after retroviral transduction of marrow-derived human osteogenic mesenchymal progenitors, *Human Gene Therapy* **8**, 1417-1427.
103. An, Y.H. and Friedman, R.J. (1999) Animal Models in Orthopaedic Research, in Y.H. An and R.J. Friedman (eds.), *Animal Models of Bone Defect Repair*, CRC Press, Boca Raton, pp. 241.
104. Buma, P., Schreurs, W. and Verdonschot, N. (2004) Skeletal tissue engineering - from in vitro studies to large animal models, *Biomaterials* **25**, 1487-1495.
105. Rzeszutek, K., Sarraf, F. and Davies, J.E. (2003) Proton pump inhibitors control osteoclastic resorption of calcium phosphate implants and stimulate increased local reparative bone growth, *Journal of Craniofacial Surgery* **14**, 301-307.
106. Rohner, D., Hutmacher, D.W., Cheng, T.K., Oberholzer, M. and Hammer, B. (2003) In vivo efficacy of bone-marrow-coated polycaprolactone scaffolds for the reconstruction of orbital defects in the pig, *Journal of Biomedical Materials Research Part B-Applied Biomaterials* **66B**, 574-580.

INDEX

223